6

W9-CID-643

Poisonous Plants of the Central United States

Poisonous
Plants
of the
Central
United States

H. A. Stephens

The Regents Press of Kansas
Lawrence

Copyright © 1980 by The Regents Press of Kansas
Printed in the United States of America
Designed by Fritz Reiber
Library of Congress Cataloging in Publication Data
. Stephens, Homer A
Poisonous plants of the central United States.

Includes bibliographies and index.
1. Poisonous plants—United States—Identification.
2. Poisonous plants—Middle West—Identification.
I. Title.
QK100.U6S73   581.6'9'0977   79-28161
ISBN 0-7006-0202-X

ISBN 0-7006-0204-6 pbk.

CONTENTS

INTRODUCTION

The purpose of this book is to provide readily accessible information about poisonous plants found in the central portion of the United States. It should be of practical use to doctors, nurses, veterinarians, ranchers, parents of small children, teachers, camp directors, field sportsmen, and leaders of outdoor youth groups. It is designed to help laymen become aware of the poisonous plants around them and identify the plants that should not be eaten, the twigs chewed, or the nectar sucked from the flowers. The rancher should be able to learn what plants on his range may be harmful to, or cause the death of, livestock.

For each species, a brief description of the plant, its habitat, and its range is accompanied by a number of photographs to aid in identification. Even then, a competent botanist should be consulted if sufficient plant material is available for an identification. The toxic principles and symptoms are given in brief form and are set apart from the plant descriptions for easy reference. The Glossary defines terms used in the book— both botanical terms and common words as they apply specifically to plants.

Botanically, the book includes only vascular plants—no bacteria, algae, or fungi, for their identification is strictly a matter for a specialist. Included are plants that cause chemical poisoning, photosensitization, and the various forms of dermatitis and hay fever, as well as those that cause mechanical injury. The family sequence and nomenclature follow that of the *Atlas of the Flora of the Great Plains* (T. M. Barkley, ed.). Following the descriptions is a table of the plants found in the central states that are known—either from chemical tests or from having caused illness—to be poisonous. These plants are grouped according to type of toxin or mechanical injury.

Geographically, the poisonous plants of the central part of

the United States are covered in one form or another. Plants of local or regional occurrence outside the central states are not listed or discussed but may be mentioned along with a closely related, central-states species. However, approximately 80 percent of the plants discussed are found throughout, or are common in other regions of, the United States. Sixty-three percent of these plants are found in the major grazing regions either in the mountains or on the plains outside the central states. Also, about 50 percent of the plants included are found across southern Canada and 15 percent in northern Mexico.

There are only a few areas in the United States where poisonous or injurious plants are absent. Everyone should be aware of this fact. Fortunately, most of these plants require the ingestion of a large amount of material before serious harm is done. Nevertheless, a few of the plants may cause death from eating only a small amount of material; these instances are noted in the text. A number of the plants listed are so distasteful that neither humans nor livestock will eat them. Other plants cause an immediate, strong, burning sensation in the mouth, and the material will be spit out.

The descriptions of the plants were made from living specimens, mostly in their habitat, but a few were grown in a garden or greenhouse. The field plants were easily located because the author has previously collected plants in all counties of the Plains states. It was merely a matter of returning to the locations where the plants were known to exist in sufficient numbers for collection and photography. This meant visiting each location at least twice a year to document all stages of growth and development. The majority of these plants, both poisonous and nonpoisonous, that have been collected by the author are on file in the herbarium of the University of Kansas.

No book of this nature can be complete and final, for new information is constantly being discovered. But an attempt has been made here to assemble the known facts in a condensed form for easy use. It would be difficult for an author to compile a book of this sort entirely from his own experience and information, for he would have to be a specialist in several fields and have spent many years in each. Therefore, the botanical information that follows is from the author's personal knowledge, gained through experience with the plants. The chemical information was drawn from various sources in the literature. If further knowledge of the history or chemistry of a poison is desired, the reader should consult John M. Kingsbury's *Poisonous Plants of the United States and Canada* (1964).

Ronald McGregor, curator of the University of Kansas herbarium, and Ted M. Barkley, curator of the Kansas State University herbarium, were generous in permitting me free use of their herbaria. William T. Barker, curator of the North Dakota State University herbarium, was kind enough to locate and send seeds that I was unable to find. Ralph Brooks, Kansas State Biological Survey, identified many of the difficult species and also furnished some of the *Crotalaria* photographs; James S. Wilson, Emporia State University, permitted me to use the two *Digitalis* flower photographs. Except for these, all other photographs were taken by the author either in the field or in the laboratory. A few of these have been published previously and are used here by permission of The Regents Press of Kansas. The staff and management of the greenhouse in Gage Park, Topeka, let me roam at will among their plants, and many of the ornamental plants were photographed there. Frederick W. Oehme, professor of surgery and medicine, College of Veterinary Medicine, Kansas State University, was kind enough to review the manuscript. To all of these people, I wish to express my thanks and appreciation.

In addition, I wish to thank the many individuals of various county, state, and federal agencies who have offered suggestions and aided the work in other ways. Their help has been valuable.

GLOSSARY

Achene. A small dry fruit with one seed and a thin wall which does not split open.

Alternate. With the leaves situated at different levels on the stem and usually on the other side of the stem.

Aril. An appendage developing from the hilum of some seeds and partly or completely surrounding the seed. It is inside the fruit wall.

Axil. The upper angle between the stem and leaf.

Bilaterally symmetrical. Used here to mean that if a flower without regular radiating petals (as in the apple flower) is cut in two lengthwise, the two halves will be identical, such as the *Digitalis* flower.

Bipinnate. A compound leaf with the blade divided into sections on each side of a central axis, each section divided again in the same manner.

Blade. The expanded, flat portion of a leaf.

Bracts. Small, leaf-like structures often directly below a flower or group of flowers.

Capsule. A dry, pod-like fruit with two or more sections developed from a compound ovary, usually splits open from top to bottom when mature.

Clavate. Gradually thickened toward the apex; club-shaped.

Cordate. Heart-shaped, with two rounded lobes at the base.

Corm. A solid, bulb-like, underground structure from which the stem arises.

Crown. The part of the plant where the stem and root join, usually at or just below the ground surface.

Decumbent. Reclining but with the outer end of the stem elevated.

Decurrent. A condition where the blade or leaf stalk is extended downward as a flange or ridge along the organ below.

Dentate. Toothed along the margin.

Disk. The circular center of a composite flower, such as a daisy.

Entire. A leaf margin without teeth.

Fascicle. A small bundle or cluster, often of bulbs or tubers.

Follicle. A dry, pod-like fruit developed from a single ovary; when dry, opening along the inner suture.

Fusiform. Thick in the middle and tapered at both ends.

Glabrous. Without hairs of any kind.

Hilum. The scar left on a seed where it breaks from attachment with the pod or fruit.

Keel. The two lower, united petals of the flowers of the legumes.

Lanceolate. Narrow but with a slightly wider base and tapering to a point.

Leguminous. A bilaterally symmetrical flower such as a bean or pea flower; in fruits, a pod similar to that of a bean or pea.

Linear. Long, narrow with parallel sides.

Node. The area where the leaf is attached to the stem.

Oblanceolate. Narrow but with the broadest part toward the tip.

Obovate. Egg-shaped with the large end toward the apex.

Opposite. With the leaves on opposite sides of the stem at the same level.

Ovary. The part of the flower which becomes the fruit.

Ovate. Egg-shaped with the broad part at the base.

Palmate. A compound leaf with the divisions (leaflets) radiating from a common center.

Panduriform. A leaf or other structure which is fiddle-shaped with the large end out.

Panicle. A loose, many-branched inflorescence with or without a central axis.

Papillate. With minute nipple-like projections.

Pappus. Usually a ring of long or short hairs or scales around one end of a fruit or seed; in some cases the ring is on a stalk.

Pedicel. The stalk of an individual flower.

Peduncle. The flower stem along which are the flowers, either with or without a stalk of their own.

Petiole. A leaf stalk, between blade and stem.

Photosensitization. A condition in which the skin becomes red and itches; eventually the affected part may die. It is caused by a hypersensitivity to light, especially if the animal has eaten foods containing certain chemicals.

Pinnate. With the blade divided into sections arranged along a central axis.

Processes. Epidermal projections, as in the coarse projections on a milkweed pod.

Raceme. With the flowers arranged singly on a central axis but each flower on a stalk.

Ray. The strap-shaped "petal" of a composite flower.

Regular. The parts of a flower arranged radially from a common center.

Rhizome. A horizontal stem usually underground, used for vegetative reproduction.

Rugose. Rough and minutely wrinkled.

Sector-shaped. The shape of a section of an orange.

Sessile. Without a stalk, attached directly to another plant part.

Silique. A slender seed "pod" of the mustard family. At maturity it splits open along the sides.

Simple. With the leaf blade in one part, not divided.

Sinuate. With the margin wavy in and out, not deep enough to be lobed.

Spathe. A large, solitary bract surrounding certain flowers.

Spike. An inflorescence with a central axis, the flowers attached to it without a stalk.

Stipule. A small, leaf-like structure just below the attachment of the leaf with the stem.

Striate. Minutely ridged with either a solid ridge or one that is minutely papillate.

Terete. Circular in cross section.

Tuberculate. With small projections, larger than papillae, the line of demarcation being rather indefinite.

Umbel. A cluster of flowers or fruits whose individual stalks radiate from a common center on a greater peduncle or stem.

Undulate. With the margin of a leaf wavy vertical to the plane of the blade, as if the margin were too large for the center.

Vascular plant. A plant having a system of tubes for the distribution of water, food materials, and foods.

Poisonous
Plants
of the
Central
United States

Pteridium aquilinum (L.) Kuhn
Bracken fern, brake

POLYPODIACEAE

Herbaceous perennial fern from an underground, dark brown branching rhizome extending for several meters. Leaves directly from the rhizome, 5–15 dm high, 2–3 times divided; blade broadly triangular in outline, 3–7 dm long, ultimate divisions 2–3 cm long, 1 cm wide, margin rolled under. Reproductive spores produced beneath the rolled edge in August or September.

Woods or open areas, often in dense stands in newly cleared land; not in open plains. Including varieties, found throughout the United States and Mexico and across southern Canada to Alaska. Common in the Black Hills and in northeastern North Dakota.

POISONING: The rhizome is the most poisonous part of the plant, but all parts are poisonous whether green or dry. Since bracken grows in wooded grazing areas and in meadows mowed for hay, it is readily available to livestock but is distasteful to most animals. When bracken is eaten by a horse, the thiaminase in it causes a deficiency of thiamin. In cattle, however, the action is quite different and the actual poisonous principle is not known. It is cumulative and the symptoms may not be evident for 2–3 weeks after ingestion, when the disease has progressed too far for cure.

SYMPTOMS: Horses develop an unsteadiness, nervousness, congestion of mucous membranes, loss of weight, constipation, and on occasions stand with their legs spread and backs arched. The symptoms in cattle are similar but include high temperature, salivation, edematous swelling, difficult breathing, and hemorrhaging— with blood in the feces and often from the nostrils.

Other species: *Onoclea sensibilis* L., sensitive fern. Herbaceous perennial from an underground rhizome; sterile leaves usually 2–5 dm high, triangular, 18–25 cm long and as wide, deeply lobed; fertile fronds 2–5 dm high, the spore-producing portion 10–12 cm long, 2–4 cm wide, becomes brown and persists through the winter. Marshes; open, wet woods; and hay meadows on low ground. Newfoundland to North Dakota, south to Texas and Florida. This plant is of doubtful danger but a few cases of horses being poisoned from it are on record. The animals showed nervous disorders and lack of coordination.

2

Pteridium aquilinum
rhizome

Pteridium aquilinum leaf

Pteridium aquilinum underside of leaf

Pteridium aquilinum leaf

Onoclea sensibilis leaves

Taxus baccata L.   **TAXACEAE**
English yew

An ornamental spreading tree or shrub usually kept trimmed as foundation planting, evergreen. Leaves closely spaced on the branches, linear, 2–3 cm long, 2–3 mm wide, flat, rigid, pointed end, dark green above, pale below; leaf stalk short, appressed against the stem. Male and female reproductive structures produced on separate plants, axillary, 2–3 mm wide, inconspicuous; May–June. Single seed produced in a cup-shaped aril that is bright red with a waxy bloom, open at the end, 9–11 mm long, 8–9 mm thick. Seed ovoid, about 7 mm long and 4.5 mm thick, red-brown, smooth.

Cultivated in the southern half of the United States as far north as Virginia and Kansas.

Other species: *T. cuspidata* Sieb. & Zucc., Japanese yew, very similar but more hardy and probably more commonly planted. Two native species: *T. brevifolia* Nutt. on the West Coast and *T. canadensis* Marsh. in the East from North Carolina and Kentucky to Newfoundland and Manitoba. Both are equally dangerous.

POISONING: The poisonous parts are the green or dry foliage, bark, and seeds. The poisonous principle is taxine, an alkaloid. Children have been poisoned from eating the seeds (the red aril is edible) or chewing the leaves. Cattle may be poisoned from eating the native forms or the trimmings from cultivated plants. Frequently large quantities will be eaten and death comes quickly, so rapidly in fact that dead cattle have been found near the plants and with foliage still in their mouths.

SYMPTOMS: Human symptoms are weakness, trembling, diarrhea, vomiting, slow heartbeat, difficult breathing, convulsions, and coma. If large quantities have been eaten, these symptoms develop rapidly and death results. In livestock, the symptoms are slowing of the heart, nervousness, difficult breathing, gastroenteritis, diarrhea, and collapse.

Taxus baccata bush

Taxus baccata leaves

4   *Taxus baccata* flower bud

Taxus baccata seed

Taxus baccata arils

Aconitum columbianum Nutt.   RANUNCULACEAE
Monkshood

Herbaceous perennial 3-6 dm high from a thickened root. Stem finely hairy, usually simple or branched at the inflorescence. Leaves alternate, drooping; leaf stalks nearly as long as the blade; blade broadly ovate in outline, 8-15 cm across, deeply palmately divided into 3-5 segments, each of which is incised and coarsely toothed. Flowers in terminal racemes about half the total height of the plant; flowers bilaterally symmetrical, about 25 mm long, blue or purple, the upper sepal forming a "hood" over the petals; July-August. Fruit a cluster of 3-4 erect follicles 10-15 mm long, close together when green but spreading at maturity; many seeds. Seeds irregular, 3-4 mm long, 1.5-2 mm wide, brown, with deep, flangelike wrinkles on the surface.

Moist areas along creeks in open woodlands or in mountain valleys. New Mexico to Canada and west to California; common in the Black Hills.

Other species: *A. napellus* L., the garden monkshood. A similar plant found in gardens, especially in the eastern United States.

POISONING: All parts of the plant contain a number of alkaloids, the best known being aconitine, and death may occur within a few hours after ingestion. The root might be mistaken for other roots which are edible, but monkshood root produces a tingling sensation in the mouth. The leaves seem to be most poisonous just before flowering time. *Delphinium* is similar but its flower has a spur instead of a "hood." The poisons are similar.

SYMPTOMS: Human symptoms consist of irregularity of heartbeat and breathing, nausea, loss of speech control, dizziness, chest pains, and tingling of the skin. In livestock, symptoms are salivation, frequent swallowing, bloating, and prostration.

5

Aconitum columbianum leaves

Aconitum columbianum plant

Aconitum columbianum fruits

Aconitum columbianum flowers

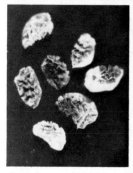

Aconitum columbianum
seeds

Aconitum columbianum
root

6

Actaea rubra (Ait.) Willd.   **RANUNCULACEAE**
Baneberry

Herbaceous perennial 5-7 dm high from a thickened rhizome. Leaves alternate, broadly ovate in outline, 2-4 dm across, divided into 3 main divisions, each divided again, the ultimate leaflets lobed and toothed, 3-5 cm long. Flowers in a short, thick raceme on a long peduncle at the end of the stem; each flower hemispherical in outline, the stamens radiating from the center, the petals inconspicuous, all parts white; flower stalks slender; May–June. Fruits globose 9-11 mm diameter, smooth, shiny, white or bright red (which makes them attractive to children), 10-12 seeds. Seeds sector-shaped, red-brown, 3-4 mm long, surface pitted.

Rich, moist soil in woodlands, quite shade tolerant. Across southern Canada and the northern United States from New England to Oregon and as far south as Indiana, Iowa, and Colorado. Common in the Black Hills, northeastern South Dakota, and wet wooded areas of North Dakota.

Other species: *A. alba* (L.) Mill., white baneberry. Enters our area in northeastern Kansas and southeastern Nebraska. It is similar to the white-fruited form of *A. rubra,* but the fruit stalks are coarse and thick. Nova Scotia to Manitoba and Minnesota, south to eastern Oklahoma, Louisiana, and Georgia.

POISONING: All parts of the plant contain the irritant oil protoanemonin, a glycoside. The roots and berries are the most poisonous. No deaths of either humans or livestock are on record for the United States, but a few berries will bring illness to people.

SYMPTOMS: Gastroenteritis, stomach cramps, headache, dizziness, vomiting, diarrhea, and circulatory failure.

Actaea rubra plant

Actaea rubra white form fruit

7

Actaea alba plant
Actaea rubra red form fruit

Actaea alba fruit
Actaea alba flower

Actaea alba seed

8   *Actaea alba* root

Delphinium virescens Nutt.   RANUNCULACEAE
Larkspur

Perennial to 1 m high, from a taproot or a fascicle of small tuberous roots. Stem simple or branched, hairy; leaves mostly below the middle. Leaves alternate, circular in outline, 5–10 cm across, deeply dissected into numerous linear lobes 2–5 mm wide, larger leaves near the base, smaller ones above; leaf stalk about as long as the blade. Flower racemes terminal on short axillary branches, the raceme elongating while in flower to nearly half the length of the main stem, flower stalk 5–15 mm long, ascending at the base and curved outward near the flower. Flowers bilaterally symmetrical, 2–3 cm long including the spur, white tinged with purple, spur ascending; May–June. Fruit a cluster of 3–5 erect follicles 15–22 mm long, remaining erect and appressed to the stem when mature, thin-walled, opening at the top; many seeds. Seeds irregular, about 2 mm long, 1.25 mm wide, dark brown, covered with ridge-like projections.

Open prairies, pastures, roadsides, either sandy or rocky limestone soil. Wisconsin and Manitoba, south in the Great Plains to Texas and Louisiana.

Other species: _D. tricorne_ Michx., dwarf larkspur. A shorter, stockier plant, the leaf segments wider, the flowers dark blue, the flower stalks spreading, the follicles curving outward from each other; the habitat is often more in open woods. Pennsylvania to Minnesota, south to Oklahoma and Georgia. In addition, there are numerous cultivated varieties all of which are toxic.

POISONING: Alkaloids such as delphinine and ajacine are the poisonous principles. Cattle are the main animals subject to poisoning; horses will eat the plant but usually not in harmful amounts; sheep are the least affected and have been used in larkspur control on the range. The entire plant is toxic including the roots and seeds. It is especially dangerous in early spring because the rosette of leaves appears before other forage is abundant; also, the plant becomes less toxic as it matures. All species of _Delphinium_ should be considered dangerous even though they are not all equally poisonous.

SYMPTOMS: Restlessness, stiffness, muscular twitching, rapid and difficult breathing, bloating, nausea, constipation, rapid pulse, and general bodily weakness. An animal may stand with feet wide apart and suddenly fall, later to rise and stand for a few minutes before falling again. Death occurs from clogging, or paralysis, of the respiratory system.

Delphinium virescens leaf

Delphinium virescens
plant

Delphinium virescens
flower

Delphinium virescens flower

Delphinium virescens fruit

Delphinium virescens
seeds
   Delphinium virescens
   root

Delphinium tricorne plant

Delphinium tricorne leaf

Delphinium tricorne seed

Delphinium tricorne fruit

11

Ranunculus sceleratus L.   RANUNCULACEAE
Cursed crowfoot

An erect annual, 2–5 dm tall, glabrous, hollow stem with many branches. Basal leaves kidney-shaped, deeply lobed, 3–6 cm long, 4–7 cm wide; upper leaves of 3 linear lobes or often a single lobe. Flowers grouped at the top of the plant; flower stalks 1–3 cm long; flowers regular, the 5 petals 2–3 mm long, light yellow; May–June. Achenes in a short, cylindric head, 5–10 mm long; achenes flattened, circular, 1 mm across, yellow, beakless or nearly so.

Marshes, wet ditches, meadows, stream banks, and lake shores. Newfoundland to Alaska and south over most of the United States.

Other species: *R. abortivus* L., wood buttercup. Similar to *R. sceleratus* but the basal leaves are not lobed and the achenes have a short beak. Moist, shaded areas. Labrador to Alaska, south to Colorado, Texas, and Florida. *R. cymbalaria* Pursh, shore buttercup. Stoloniferous, creeping perennial, 5–10 cm high, petals 3–5 mm long. Mud of slow streams or lake shores and in grassy marshes. Throughout most of the United States and southern Canada, also in Asia. There are several other species of buttercup in the central states and all should be regarded with suspicion. The plants should be handled carefully because the juice causes small blisters on the skin of some people.

POISONING: Protoanemonin, a glycoside, is the poisonous principle in buttercups and may be found in all above-ground parts. It is not a strong poison, and relatively large amounts of the plant must be consumed before livestock show signs of illness.

SYMPTOMS: Salivation, gastric pains accompanied by a loss of appetite, colic, diarrhea, and slowing of the heartbeat. The milk from affected cows is bitter.

Ranunculus sceleratus plant

Ranunculus sceleratus leaf

Ranunculus sceleratus
achene

Ranunculus sceleratus
flower

Ranunculus abortivus leaf
Ranunculus sceleratus flower, fruit

Ranunculus cymbalaria achene

Ranunculus cymbalaria plant   13

Podophyllum peltatum L.
Mayapple, mandrake

BERBERIDACEAE

Herbaceous perennial 4-6 dm high from a coarse, creeping rhizome about 5 mm diameter, white or brown. Stems that do not have flowers have one rounded leaf deeply dissected into 5-9 lobes, 2-4 dm across, the stalk attached in the center; flowering stems have 2 lobed leaves, 2-4 dm across, each leaf stalk 15-25 cm long, attached near the blade margin. Flowers single in the axil of the two leaf stalks, regular, 3.5-5 cm across, 6 white petals; flower stalks 2-3 cm long, curved so that the flower faces downward; April-May. Fruit ovoid, 4-5 cm long, smooth, yellow when ripe. Seeds sector-shaped, 6-7 mm long, 3-3.5 mm across, pale brown, minutely and finely wrinkled.

Moist, rich woods and wooded hillsides bordering streams; often in large stands in recently cleared land. Quebec to Minnesota, south through eastern Nebraska and Kansas to Texas and Florida.

POISONING: All parts of the plant, especially the rootstock, young shoots, and unripe fruit contain the resinoid podophyllin. Ripe fruits are pleasant tasting to many people and may be eaten without danger in small quantities. Podophyllin has been used as a drug and most cases of human poisoning have been from an overdose of the drug. Livestock may be poisoned from eating the leaves or unripe fruit but usually it is not severe.

SYMPTOMS: In humans, gastroenteritis and vomiting. In livestock, watering of the eyes and mouth, diarrhea, and excitement.

Podophyllum peltatum
fruit

Podophyllum peltatum
single leaf plant

14

Podophyllum peltatum
2-leaf plant

Podophyllum peltatum
flower

Podophyllum peltatum
flower

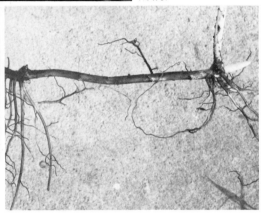

Podophyllum peltatum
seed
  Podophyllum peltatum
  rhizome

Menispermum canadense L.   MENISPERMACEAE
Moonseed

A twining, woody vine to a length of 8 m, often several stems twisted together, the ends dying back during the winter. Leaves alternate, circular in outline, 7–13 cm across, margin with 5–7 rounded angles or shallow lobes; leaf stalk 6–10 cm long, attached to the blade 2–4 mm from the margin, veins arranged palmately. Flowers axillary,   15 in drooping panicles 4–4.5 cm long, each flower 4–8 mm across, greenish white; May–June. Fruits in grape-like clusters, each fruit globose, 6–9 mm diameter, dark blue with a whitish bloom resembling grapes, 1 seed in each fruit. Seeds flat, circular with a missing wedge, 6.5–8 mm across, center smooth surrounded by a rough ridge and a marginal ridge.

Wooded areas, fences near woods, or climbing over bushes along stream banks. Quebec to Manitoba, south to Oklahoma, Arkansas, and Georgia. Common in eastern Kansas and along the Missouri River in Nebraska.

POISONING: All parts of the plant contain an alkaloid which causes illness. The rootstock is apparently the most poisonous, but is also most inaccessible. Children have reportedly been poisoned from eating the fruits and livestock from eating the foliage and probably the fruits. However, the foliage is bitter and seldom eaten.

SYMPTOMS: Gastroenteritis.

Menispermum canadense leaf

Menispermum canadense flower

Menispermum canadense fruit

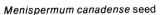
Menispermum canadense seed

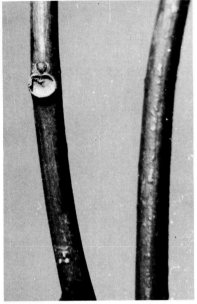

Menispermum canadense stems
Menispermum canadense stems

Dicentra cucullaria (L.) Bernh.   **FUMARIACEAE**
Dutchman's breeches

Herbaceous perennial with leaves and flower stems directly from a cluster of small, underground tubers; tubers pink with fibrous roots below, the whole cluster 2-3 cm across. Leaf stalk 8-15 cm long; blade broadly ovate in outline, 8-12 cm wide, deeply dissected, usually with 3 main divisions each divided into narrow segments 1-2 mm wide. Main flower stem 10-20 cm long, the raceme up to 10 cm long with 3-8 flowers on slender, recurved stalks; 1-15 flower stems arising from a cluster of tubers. Flowers irregular, resembling a pair of flare-legged Dutchman's breeches hanging top down, 12-18 mm long and as wide, white or pink; April-May. Fruit a pod extended into an acuminate tip, several seeds. Seeds nearly circular with an indentation at the point of attachment, about 2 mm across and 1 mm thick, glossy black.

Rich, moist soil of woodland areas, along rock ledges in woods, or along stream banks. Quebec to North Dakota, south to Oklahoma, Arkansas, and Georgia.

17

Other species: *D. spectabilis* Lem., the common garden bleeding heart. A close relative and equally dangerous.

POISONING: All parts of the plant, especially the tubers, contain alkaloids such as protopine, protoberberine, and aporphine. Poisoning is rare in either humans or livestock because the plant is not common and does not grow in areas where children play or livestock graze.

SYMPTOMS: Trembling, staggering, holding the head high, salivation, difficult breathing, and convulsions.

Dicentra cucullaria plant

Dicentra cucullaria leaf

Dicentra cucullaria flowers

Dicentra cucullaria roots

Dicentra cucullaria fruits

Dicentra cucullaria seeds

Cannabis sativa L.
Marijuana, hemp

CANNABACEAE

Herbaceous annual 1-2.5 m high, a central stem with many branches; stems rough. Leaves opposite below and alternate above; palmately compound with 3-7 lanceolate, toothed leaflets 10-12 cm long, 1 cm wide; leaves of the inflorescence usually simple and lanceolate; leaf stalks about half as long as the blade. Male and female flowers on separate plants toward the end of a branch; male flowers in a loose, leafy panicle in leaf axils; female flowers in short, dense panicles in leaf axils; flowers 5-7 mm across, greenish white; flower stalks slender, 3-4 mm long; July-October. Achene lenticular, 3-4 mm long, 2.5 mm wide, slightly ridged or keeled on two angles, brown, smooth, enclosed in bracts.

Weedy areas in pastures, along draws, fence rows, old feed yards, or brushy stream banks. Common along the east edge of Kansas and Nebraska, extending into the sandhills of central Nebraska. Introduced from Asia and now widespread throughout the United States except in desert regions.

POISONING: All parts of the plant contain the resin tetrahydrocannabinol. Death is not common in either humans or livestock; but in countries where death has resulted, it was caused by the narcotic effect on the heart. Pollen from the flowers is a major cause of hay fever for people in some sections of the central states.

SYMPTOMS: In humans, after ingesting or smoking the resin, the initial result is a sense of elation and well-being. This is followed by hallucinations, inability to think clearly, blurred vision, and eventual coma. In livestock, difficult breathing, trembling; sweating occurs even though the body temperature is low.

Cannabis sativa plant

Cannabis sativa leaf

19

Cannabis sativa flowers

Cannabis sativa leaf

Cannabis sativa seeds

Cannabis sativa flowers

Cannabis sativa fruits

Cannabis sativa fruits

Urtica dioica L. ssp. *gracilis* (Ait.) Seland.   URTICACEAE
Stinging nettle

Herbaceous perennial, colony forming. Stem fibrous, usually un-branched, 1–2 m high, erect, covered with stinging bristles. Leaves opposite, ovate to lance-ovate, 6–13 cm long, 4–6 cm wide, base sub-cordate to obtuse, tip long tapered; margin toothed. Flowers clustered in axillary panicles 4–7 cm long on the upper portion of the stem; each flower 1–2 mm across, yellow-green; July–August. Calyx enlarges to enclose the achenes which are ovate, 1 mm long, 0.75 mm wide, surfaces rounded, yellow.

Other species: *Laportea canadensis* (L.) Wedd., wood nettle, a related species. It is a shorter plant, alternate leaves, broadly ovate blade 10–15 cm long, 7–10 cm wide; it has similar stinging bristles. Found in moist, wooded areas along the eastern edge of the central states. *Tragia* spp., a short plant of the prairies or open woods, the leaves lanceolate and toothed. It is a member of the *Euphorbiaceae* but is listed here because it has the same type of stinging bristle as *Urtica* and the chemicals are quite similar. The effect on the skin is often more severe than that from *Urtica.*

POISONING: The plants have stinging bristles similar to a hypodermic needle with a sharp point, a capillary tube through the bristle and a sac of fluid at the base. The bristle bends as it penetrates the skin and puts sufficient pressure on the sac to force the fluid into the skin. The chemistry of the fluid is not fully known.

SYMPTOMS: In humans and livestock, itching, burning skin, often with welts. May last for an hour or more.

Urtica dioica leaf

Urtica dioica plant

21

Urtica dioica flowers

Urtica dioica stinging hairs

Urtica dioica seeds

Urtica dioica fruits

Urtica dioica flowers

Quercus spp.
Oaks

FAGACEAE

Oak poisoning of livestock is seldom reported in the central states but is quite common in the "scrub" oak areas of the Southwest where shinnery oak, *Quercus havardii* Rydb., and Gambel's oak, *Quercus gambelii* Nutt., are abundant. In those areas the trees are low and livestock can reach the young shoots. This usually occurs early in the season when other forage is limited. Only in the southern part of the central states are there sufficient numbers of oaks to cause serious harm to animals. Although a number of species are present only two

22

of them usually have branches within reach of the animals, Q. *marilandica* Muenchh., a red oak, and Q. *stellata* Wang., a white oak.

Q. *marilandica,* blackjack oak. A small tree to 8 m, scrubby in appearance, nearly black trunk and limbs. Leaves wider at the outer end, usually 3-lobed, a short bristle at the end of main veins. Acorn cup 14–18 mm wide, enclosing about one-third of the nut, cup scales coarse. Often grows in dense clusters or larger stands in sandstone soil. Massachusetts south to Florida, west to Iowa, Kansas, and Texas.

Q. *stellata,* post oak. A tree to 18 m, the young trees with low branches. Leaves with 3–5 irregular lobes, the lobes broad and rounded. Acorn cup 11–14 mm wide, enclosing about one-third of the nut, scales fine. Often in pure stands or mixed with blackjack oak, the two oaks having the same range.

POISONING: Tannin is the principal poison, but other compounds such as quercitrin and quercitin may also have some effect. Poisoning occurs only when an animal eats a large quantity of the acorns, young sprouts, or dried leaves without other forage. The young sprouts are considered beneficial in small amounts.

SYMPTOMS: Anorexia, constipation with dry, pelleted feces, followed by bloody diarrhea, rough coat, dry muzzle, thirst, weak pulse, and frequent urination.

Quercus marilandica flowers, young leaves

Quercus marilandica leaf

Quercus marilandica trunk

23

Quercus stellata fruit
Quercus marilandica fruit

Quercus macrocarpa fruit

Quercus stellata leaf

24

Quercus muhlenbergii young leaves

Phytolacca americana L. PHYTOLACCACEAE
Pokeberry, pokeweed, pigeon berry, inkberry

Herbaceous perennial from a stout, vertical rootstock 5-8 dm long, 8-12 cm thick, tapering at the bottom to coarse roots. Plant 1.5-2.5 m high, erect, branching; stem hollow, glabrous, purplish. Leaves alternate, narrowly ovate, 15-30 cm long, 6-10 cm wide, margin entire, often wavy, tapered base, glabrous; leaf stalk 1-4 cm long. Flowers in nodding racemes 7-12 cm long, loosely flowered; flowers regular, 5-7 mm across, white, tinged with purple; July–October. Fruiting racemes drooping, 10-15 cm long; berries flattened globose, 11-14 mm diameter, dark purple, juicy, many seeds. Seeds nearly circular, 2.5-3.5 mm across, flattened, glossy black.

Waste areas, roadsides, creek banks, old farmsteads, usually in moist soil. Maine across southern Canada to Minnesota, south to eastern Nebraska, Texas, and Florida.

POISONING: The exact nature of the poison is unknown but there is evidence that it is a saponic glycoside. The rootstock is the most poisonous, but all parts of the mature plant contain some poison. People often use the young sprouts as greens, but the water in which they are boiled must be thrown away. Although the fruits occasionally are used for pies, extreme care should be taken because children have been fatally poisoned from eating the fresh berries or making them into a drink to simulate grape juice. Cattle and horses rarely eat the leaves, but cases of poisoning are on record. Pigs will root out the rootstocks and have been fatally poisoned from eating them.

SYMPTOMS: In humans and livestock, severe gastric-intestinal pain and cramps, spasms, ulcerative gastritis, vomiting, diarrhea, and convulsions. Death is caused by paralysis of the respiratory muscles.

Phytolacca americana leaf

Phytolacca americana plant

25

Phytolacca americana
flowers
Phytolacca americana flowers

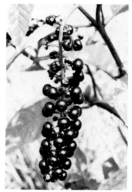

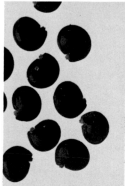

Phytolacca americana
fruits

Phytolacca americana
seeds

Phytolacca americana
root

Agrostemma githago L.
Corn cockle

CARYOPHYLLACEAE

Annual, herbaceous weed 40–75 cm high. Stem hairy, simple, branched at the inflorescence or with branches clustered near the base. Leaves opposite, the lower ones oblanceolate, 10–13 cm long, 10–15 mm wide near the outer end; upper leaves linear, 5–10 cm long, 3–6 mm wide, hairy with long silky hairs along the entire margin; leaf base slightly clasping. Flowers axillary or terminal, single on the end of a stalk 10–15 cm long; calyx ovoid, hairy, coarsely 10-ribbed, 5 linear lobes 25–35 mm long, longer than the petals. Flowers regular, 3–4 cm across, petals 5, red, obovate, notched at the apex; early summer. Capsule 12–18 mm long, opening at the end, many seeds, calyx lobes persistent. Seeds irregular, dark brown, 3–4 mm across, 2–3 mm thick, surface rough with rows of minute papillae.

26

Introduced from Europe, now widespread but not common, mainly in the northern part of the United States and southern Canada. Horticultural strains are used in flower gardens.

POISONING: The seeds contain a high percentage of githagenin and sapogenin. Human poisoning is rare, but when home-ground flour was used several cases of poisoning resulted from using wheat contaminated with corn cockle seeds. It is dangerous to feed screenings to livestock if corn cockle is known to have grown in the field. Chickens will reject the seeds or even refuse to eat a mash containing them, but cases of poisoning have been reported. Horses and cattle have been fatally poisoned from eating the plant in fruit. There is one case on record of pigs being killed from eating the roots.

SYMPTOMS: In chickens, listlessness, ruffed feathers, and diarrhea. In other animals, including humans, gastroenteritis, vomiting, dizziness, diarrhea, rapid breathing, and coma.

Agrostemma githago seeds

Agrostemma githago young plant

Agrostemma githago leaves

27

Saponaria officinalis L.
Bouncing bet, soapwort

CARYOPHYLLACEAE

Herbaceous perennial forming colonies by thick, woody rhizomes. Stems 3-8 dm high, simple or branched, enlarged at the nodes. Leaves opposite, elliptic to lance-elliptic, 7-10 cm long, 2-4 cm wide, tapered to a broad, short stalk, usually 3 main longitudinal veins. Flowers in either a compact, globose cluster or an elongated, open inflorescence; calyx cylindric, 1.5-2 cm long; corolla 2.5-3 cm across, petals 5-6, pink or white, notched at the apex, occasionally the flowers are "double"; June–October. Fruiting capsule narrowly conic or cylindric, 18-22 mm long, 5-6 mm thick, many seeds. Seeds nearly circular, about 2 mm across, 0.5 mm thick, black, minutely "bumpy," the tiny bumps often in concentric circles.

A weedy plant of unused areas, roadsides, persisting for years around old farmsteads. Introduced from Europe and now widespread in the United States.

POISONING: All parts of the plant contain saponin, a glycoside, with the higher concentration in the seeds. The plant is rarely eaten by livestock because it is distasteful to them. Often grain screenings contain the seeds, or those of *S. vaccaria* L., cow cockle, and care should be taken in feeding livestock such screenings.

SYMPTOMS: Nausea, vomiting, diarrhea, unsteadiness, irritation of the digestive system, rapid breathing, and coma.

Saponaria officinalis plant

Saponaria officinalis leaves

Saponaria officinalis flowers

Saponaria officinalis flowers

Saponaria officinalis flowers

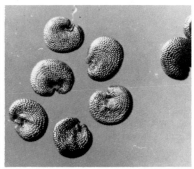

Saponaria officinalis fruits

Saponaria officinalis seeds

Sarcobatus vermiculatus (Hook.) Torr.   CHENOPODIACEAE
Greasewood

    A dense, woody, spiny shrub 1–2 m high and as broad, with many fine, yellow-gray branches. Leaves alternate, simple, linear, 15–25 mm long, succulent, flattened or terete, remain on the plant until early winter. Male flowers cone-like, 1.5–3 cm long, terminal on twigs; female flowers inconspicuous in leaf axils below the male flowers, the ovary surrounded by a cup-like calyx which enlarges in fruit; June–August. Fruit ovoid, 4–5 mm long, surrounded at the middle by a wing extending at right angles to the long axis of the seed; wing circular, 8–12 mm across, pale brown, lobed or undulate margin, papery but tough and rigid when dry.

    Flat flood plains or prairies, eroded or grassy hillsides, in alkaline soils. North Dakota to British Columbia south to California, Mexico, New Mexico, Colorado, and western Nebraska.

29

POISONING: The leaves contain soluble oxalates and are the most toxic part of the plant. The oxalate content increases as the leaves mature. Cattle and horses seldom eat the plant but sheep poisoning is quite common. Greasewood is considered useful as forage but other kinds of feed must be ingested with it to prevent poisoning. Hungry sheep should never be placed in a pasture where greasewood is common.

SYMPTOMS: Listlessness, weakness, prostration, kidney lesions, depression, weakened respiration and heartbeat, finally coma and death.

Sarcobatus vermiculatus shrub

Sarcobatus vermiculatus leaves

Sarcobatus vermiculatus branch

Sarcobatus vermiculatus flowers

Sarcobatus vermiculatus fruits

Sarcobatus vermiculatus fruits

Amaranthus retroflexus L.   AMARANTHACEAE
Pigweed, redroot

A stout annual weed 4–10 dm high with a central stem and few to many branches; stem coarse, covered with short, rather bristly hairs. Taproot stout, usually red. Leaves alternate, ovate, 7–10 cm long, 3–6 cm wide, base acute, margin undulate, finely toothed or entire; leaf stalk nearly as long as the blade. Flowers in terminal, dense panicles and in smaller panicles in the axils of the upper leaves; male and female flowers separate but on the same plant, the bracts with a short awn tip; each flower inconspicuous, greenish; June–September. Fruit a small, ovoid, thin-walled structure containing one seed. Seeds compressed ovate, about 1 mm long, glossy black.

Disturbed soil of cultivated fields, farm lots, roadsides, or in eroded areas in pastures. Introduced from tropical America and now widespread in the United States.

Other species: One other species is commonly mentioned in poisonous plant literature, *A. palmeri* Wats. Similar but with slightly more slender branches; leaves narrower, the leaf stalk often longer than the blade; flowering spike up to 6 dm long, spiny. Commonly   31 found in sandy soil. Mexico, California, and Arizona, west to Texas and southwestern Kansas.

POISONING: All parts of the plant may accumulate large quantities of nitrate. Animals may be poisoned from eating either the fresh plant or hay or silage containing the dried plant. The main animals affected are pigs, cattle, and sheep.

SYMPTOMS: Brown discoloration of nonpigmented skin, weakness, lack of coordination, trembling, paralysis of hind limbs, and abortion in pregnant animals.

Amaranthus retroflexus seeds

Amaranthus retroflexus plant

Amaranthus retroflexus leaf

32   *Amaranthus retroflexus* flowers

Amaranthus palmeri leaf

Amaranthus palmeri
flowers

Amaranthus palmeri
flowers

***Rheum rhaponticum* L.**　　　　　　　**POLYGONACEAE**
Rhubarb

Herbaceous perennial from a heavy rootstock. Flowering stem 8–12 dm high, hollow, 2–5 small leaves, a brown sheath around the stem at each leaf. Principal leaves directly from the ground; blade heart-shaped, 2.5–4.5 dm long, 2–4 dm wide, margin wavy, 5 principal veins from the leaf base; leaf stalk stout, 2–5 dm long. Flowering panicles terminal or axillary, crowded; flowers regular, 3–5 mm across, 6 petaloid divisions, greenish white, clustered; April–May. Fruit stalk 5–8 mm long, drooping; fruit 3-winged, 10–12 mm long, brown, dry. A common plant in gardens throughout the United States, especially in the north.

33

POISONING: The leaf blade contains oxalic acid and soluble oxalates, at times to such an extent that a small portion of the blade causes poisoning. The stalk of the leaf is edible and commonly used in pies and jams, but the blade must be disposed of so that livestock will not get it. People have been poisoned from eating the cooked leaf blades, and farm animals are often poisoned when the discarded blades are thrown within their reach.

SYMPTOMS: In humans, weakness, stomach pains, vomiting, irritation in the mouth and throat, and in severe cases coma and death. In livestock, weakness, salivation, staggering, convulsions, and death.

Rheum rhaponticum plant

Rheum rhaponticum flower stalk

Rheum rhaponticum flowers

34

Rheum rhaponticum
fruits

Rheum rhaponticum
flowers

Rheum rhaponticum
young fruits

Hypericum perforatum L.   HYPERICACEAE
St. Johnswort

Herbaceous perennial 4–7 dm high from a woody rootstock; often many erect stems from one rootstock, branches opposite, the lower ones without flowers. Leaves opposite, elliptic, 2–4 cm long, 1–2 cm wide, much smaller on the branches; leaves sessile, covered with minute glandular dots. Flowers in branched heads at the end of the stem and branches; 5 yellow petals 8–10 mm long with black dots on the margin; June–August. Capsules ovoid, 4–6 mm long, many seeds in a sticky substance. Seeds oblong, 0.75–1 mm long, brown, finely pitted.

Dry prairies, roadsides, ditches, open woods, or low, moist ground. Introduced from Europe and now widespread in the United States and southern Canada.

POISONING: The hypericin found in the dots of either fresh or dry foliage is a fluorescent pigment and causes photosensitization in the animal. Hypericin is not destroyed by drying or in digestion and absorption; thus, it reaches the skin through the capillary blood vessels. It shows only in light-skinned animals or in the light skin of a spotted animal. Sheep may be affected after they have been sheared. In humans, contact with the plant may cause dermatitis. Any species of *Hypericum* with glandular dots should be regarded with suspicion.

SYMPTOMS: In livestock, redness of the skin accompanied by itching, inflammation of the eyes, nose, and mouth may cause blindness, difficult breathing, or an inability to eat, thus causing death. Animals may seek relief from the itching by rubbing against objects or getting in a pond or stream. However, cold water intensifies the effect and the animal may go into convulsions.

Hypericum perforatum
plant

Hypericum perforatum leaf

Hypercium perforatum
flowers

Hypericum perforatum flowers

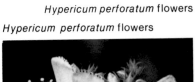

Hypericum perforatum flowers

36

Hypericum perforatum
seeds
Hypericum perforatum
fruits

Brassica hirta Moench   BRASSICACEAE
White mustard

Herbaceous annual 2–6 dm high, bristly hairy throughout, branches usually above the middle. Leaves alternate, obovate or lyrate, 10–15 cm long, 6–8 cm wide, lower leaves much larger than the upper; margin lobed or coarsely toothed. Flower racemes terminal on branches; racemes short in flower, elongating in fruit; flower stalk 1 cm long, spreading; petals 4, yellow, obovate, 1 cm long; May–June. Fruit a slender silique, 15–25 mm long with a terminal beak nearly equaling the body length, hairy, 8–15 seeds. Seeds globose, 1.5–2 mm diameter, brown, smooth.

Other species: *B. kaber* (DC.) Wheeler, charlock. Similar to *B. hirta* but less hairy, with the beak of the fruit about half as long as the body. *B. nigra* (L.) Koch, black mustard. Much taller plant, hairy below and glabrous above, flowers 1 cm across, fruit glabrous, beak short. Most common source of table mustard. *B. juncea* (L.) Coss., brown mustard. Similar to *B. hirta* but slightly taller and glabrous, lower leaves deeply lobed. *B. campestris* L., field mustard. Similar to *B. nigra* but smaller; lower leaves without stalks, upper leaves sessile and clasping the stem.

All of these species are found in waste places, roadsides and fields, or in cultivated land. All are scattered throughout the United States. Among the cultivated plants belonging to this group are: rape, horseradish, cabbage, kale, broccoli, and turnips.

POISONING: The poisonous principles vary a trifle in this group, but they all come under the class of mustard oil glycosides. All parts of the plant may be poisonous, and press cake made from the seeds should be carefully checked.

SYMPTOMS: The symptoms also vary depending on the species, the part and amount eaten, and the type of animal. In general, for both   37 humans and livestock the symptoms include gastroenteritis with severe pain, irritation of the mouth, salivation, diarrhea, difficult breathing, and in severe cases paralysis of heart and lungs. Some species are also goitrogenic.

Brassica hirta flower
Brassica hirta plant

Brassica hirta flower
Brassica kaber plant

Brassica kaber fruit

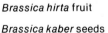
Brassica hirta fruit

Brassica kaber seeds

Descurainia pinnata (Walt.) Britt.   BRASSICACEAE
var. *brachycarpa* (Richards.) Fern.
Tansy mustard

Herbaceous annual 2-7 dm high, usually branched above the middle, finely hairy or glabrous, stem from an early rosette which is gone by maturity. Leaves alternate, 6-15 cm long, once or twice deeply dissected into narrow segments 1 mm wide, the upper leaves smaller, the basal rosette leaves with wider segments. Flower racemes terminal on branches, elongating during flowering to 10-25 cm by fruiting time; flowers regular, 2-3 mm across, 4 yellow petals; flower stalks short at flowering, elongating to 10-15 mm in fruit; April-June. Fruit cylindric to clavate, 7-12 mm long, 1-2 mm thick, indented between the seeds; fruit stalks spreading, fruit ascending. Seeds oblong, about 1 mm long, less than one-half as wide, golden brown, minutely striate with rows of rounded projections.

Fields, roadsides, unused areas, and barren spots in pastures, in rich or poor soil. Widespread throughout the United States and southern Canada.

Other species: *D. sophia* (L.) Webb. Similar but with fruits 15-25 mm long and less than 1 mm thick; more common in the western part of the Plains. *D. richardsonii* (Sweet) Schulz, a taller plant with more densely crowded racemes; northern part of the range.

POISONING: All parts of the plant are poisonous but the toxin is not definitely known. The symptoms are similar to those of selenium poisoning, but tansy mustard contains very little selenium.

SYMPTOMS: Blindness, aimless wandering, and a paralyzed tongue, making the animal unable to swallow either food or water.

Descurainia pinnata leaf

Descurainia pinnata plant

Descurainia pinnata fruits

Descurainia pinnata seeds

Descurainia sophia plant

Descurainia sophia leaves

Descurainia sophia flowers

Descurainia sophia fruits

Descurainia sophia seeds

40

Thlaspi arvense L.   **BRASSICACEAE**
Penny cress, fanweed

Herbaceous annual, simple or branched, 3–6 dm high, often several stems from one crown. Leaves alternate; basal rosette leaves oblanceolate, 7–10 cm long, 1–3 cm wide, sinuate or shallowly toothed, soon dying; stem leaves lanceolate to elliptic, 7–10 cm long, diminishing in size toward the top, irregularly toothed, base without a stalk, a free lobe extending along each side of the stem. Flowers in terminal racemes on upper branches; flowers regular, 3–4 mm across, 4 white petals; April–June. Racemes elongating in fruit to 2–3 dm. Capsules with 2 lateral wings, notched at the tip, total length 1–1.5 cm, width 10–12 mm, yellow, many-seeded. Seeds ovate, about 2 mm long, 1.25 mm wide, flattened, striate with ridges close together, dark brown.

Weedy areas, fields, ditches, and roadsides, either moist or dry. Native of Europe, now established through most of North America.

POISONING: The seeds of penny cress contain allyl isothiocyanate. Livestock may get the seeds from eating the live, mature plant, from being fed grain screenings containing large amounts of the seeds, or from eating hay containing the dried, mature plant.

SYMPTOMS: Gastric distress, diarrhea with bloody feces, colic, and in extreme cases paralysis of the heart and respiratory system.

Thlaspi arvense plant

Thlaspi arvense basal leaves

Thlaspi arvense flowers

41

Thlaspi arvense seeds

Thlaspi arvense fruits   *Thlaspi arvense* fruits

Anagallis arvensis L.
Scarlet pimpernel

PRIMULACEAE

Herbaceous annual, decumbent to erect, branches 1–2 dm long, often hidden in the grass. Stem 4-angled, glabrous. Leaves opposite, elliptic to ovate, 1–1.5 cm long, entire, sessile. Flowers single in leaf axils, regular, about 1 cm across, brick red to salmon-color, darker in the center, occasionally blue or white; the 5 petals rounded, united at the base, opening only in good weather; June–August. Fruit stalks erect, about 1 cm long, becoming 1.5 cm long and recurved at maturity. Calyx persistent in fruit. Capsules globular, 5–6 mm diameter, thin-walled, opening by splitting around the middle; many seeds. Seeds irregular, 1–1.3 mm long, 3-angled, often one side flat, the other side ridged; dark brown, the surface finely papillate.

Waste areas, lawns or pastures where it may be easily overlooked. Native of Eurasia, now scattered throughout the United States except in the mountains. The English call it "Poor man's weather glass" because the flower closes with cloudy weather.

POISONING: There is very little definite information about the poison or its effect on animals. It apparently has killed sheep and calves, and some experimental work verifies this. Some people are subject to a skin rash from contact with the leaves.

SYMPTOMS: In livestock, depression, anorexia, and diarrhea. There may be internal hemorrhaging and lung congestion.

Anagallis arvensis plant
Anagallis arvensis flower

Anagallis arvensis leaves, fruit

Anagallis arvensis opened fruits

Anagallis arvensis seeds

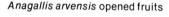

Prunus virginiana L.
Choke cherry
ROSACEAE

Woody perennial usually found in colonies, 2–5 dm high; trunk brown, branches with a purplish cast and with many horizontal, pale markings (lenticels). Leaves alternate, simple, obovate, 5–10 cm long, 3–5 cm wide, finely toothed; petiole 15–20 mm long. Flowers in axil- 43 lary racemes, 5-7 cm long, on a short, new branch; racemes densely flowered; flowers regular, 7-8 mm across, 5 white petals; April–May. Fruits globose, 8-11 mm diameter, dark red-purple, glossy, juicy; stone oval, about 7 mm long, cream-colored.

Fence rows, hedge rows, margin of woods, creek banks, fairly open areas. Range (including varieties) across southern Canada and throughout the United States except in the extreme southern areas.

Other species: *P. serotina* Ehrh., black cherry. Larger tree growing singly up to 15 m; leaves oblong or narrowly ovate, 4-10 cm long, 2-4 cm wide. Rich woods, open hillsides. Nova Scotia to Ontario and Minnesota, south to eastern Nebraska, Texas, and Mexico, east to Florida. *P. pensylvanica* L., pin cherry. Tree to 8 m, leaves elliptic to elliptic-lanceolate, 4-7 cm long, 2-3.5 cm wide. Flowers regular, 12-14 mm across, single or clustered on stalks 2 cm long. Fruit 5-8 mm diameter, red, glossy. Moist woods, hillsides, and stream banks. Newfoundland to the District of Mackenzie and south in the mountains to Colorado, from North Dakota southeast through Iowa to Tennessee and North Carolina, also in the Black Hills. *P. persica* Batch., cultivated peach. There are other species of *Prunus* either native or cultivated in our area and many should be considered toxic.

POISONING: The toxic principle is amygdalin, a cyanogenic glycoside which breaks down in hydrolysis to hydrocyanic acid. Amygdalin is present in the bark, leaves, and seeds (the kernel inside the stone). People have been poisoned by drinking a tea made from the leaves, eating the seeds, or chewing the twigs. Livestock poisoning usually comes from eating the wilted leaves which contain more cyanide than the fresh leaves. (The wilting of leaves may be caused from drought or from trimmings left where the animals may get them.) Goats, especially, peel the bark from the tree and eat it. Death may come suddenly if the animal eats the dried leaves and then drinks a quantity of water.

SYMPTOMS: In humans, difficult breathing, paralysis of the throat, spasms, and coma. In livestock, slobbering, difficult and increased rate of breathing, dizziness, rapid and weak pulse, spasms, and coma. These come in rapid succession, and death may result within an hour after ingestion of a quantity of the material.

Prunus virginiana flowers

Prunus virginiana leaves, flowers

Prunus virginiana seed

Prunus virginiana fruits

Prunus serotina leaf

Prunus serotina tree

Prunus serotina flowers

Prunus serotina flowers

45

Prunus pensylvanica
fruit

Prunus pensylvanica leaves, fruit

Prunus pensylvanica
branch

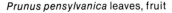

Astragalus racemosus Pursh   FABACEAE
Milk vetch, creamy poison vetch

Herbaceous perennial from a stout crown and long, heavy taproot. Stems 3–7 dm high, 5–50 stems from one crown, branching toward the top, erect or slightly decumbent, reddish and minutely hairy. Leaves alternate, pinnately compound, 6–12 cm long, 7–21 leaflets; leaflets elliptic to oval, 2–4 cm long, 4–12 mm wide, minutely appressed hairy. Flowers in racemes 8–12 cm long at the end of a branch, usually densely flowered; flower stalks 1–3 mm long; flowers leguminous, 12–15 mm long, white, pale lavender, or white with a bluish spot on the keel; May–June. Racemes elongating in fruit to 12–18 cm, fruit a leguminous pod, 3–3.5 cm long, 4–5 mm thick, angled, glabrous, drooping, several seeds. Seeds bean-shaped with the hilum near one end, 3–3.5 mm long, 1.5–2 mm wide, dark brown, smooth.

Other species: *A. bisulcatus* (Hook.) Gray, milk vetch. Similar to *A. racemosus* but with smaller leaflets, more densely flowered racemes, flowers smaller and purple; pods 1.5–2.5 cm long; stem definitely reddish. *A. pectinatus* Dougl., poison vetch. 3–5 decumbent stems from a tough crown; stems 3–5 dm long, striate, few appressed hairs. Leaflets 7–15, linear, 4–5 cm long, 1–2 mm wide, sparingly appressed hairy. Flowers cream-white, 20–25 mm long. Pods drooping, 15–20 mm long, 7–9 mm wide, sides rounded, glabrous. Seeds bean-

46

shaped, hilum near one end, 3–4 mm long, 2–2.5 mm wide, tan-brown.
A. mollissimus Torr., woolly loco. Low perennial from a narrow
taproot; leaves and flower stems directly from the crown, the whole
plant densely woolly. Leaves decumbent or erect, pinnately compound, 15–25 cm long, 15–31 leaflets; leaflets oval or broadly elliptic,
1–2.5 cm long, 6–12 mm wide. Flower stems with racemes 15–30 cm
long, raceme about half this length, flowers usually crowded. Flowers
leguminous, 15–25 mm long, dull purple with pale laterals. Raceme
elongating in fruit; fruit a leguminous pod, usually curved, 15–20 mm
long, 8–10 mm thick, glabrous. Seeds kidney-shaped to oblique heart-
shaped, 1.5–2 mm long, 1–1.5 mm wide, light brown, smooth. Ingestion of the plant produces symptoms of loco poisoning.

All four species are found in prairie areas of the Plains states.
The first three extend from the Canadian border to Texas and west
from the Great Plains into the foothills of the Rocky Mountains. *A.
bisulcatus* and *A. pectinatus* are found further west than the others.
A. mollissimus has a limited range, from northwest Nebraska south
through southeast Wyoming, Colorado, and western Kansas to Texas
and northeast New Mexico.

POISONING: Selenium is the principal poison found in the first
three species. All parts of these plants are poisonous and remain so
in drying. They grow only in soils containing selenium and may accumulate dangerous quantities of it. Other toxins may also be present. Fortunately, livestock avoid these species if other forage is
available. The poison in *A. mollissimus* has not been thoroughly identified but the symptoms are those of other loco weeds.

SYMPTOMS: In selenium poisoning, livestock suffer depression,
diarrhea, loss of hair, breakage at the base of the hoof, excessive
urination, difficult breathing, rapid and weak pulse, and coma. Death
results from the failure of the lungs and heart. In *A. mollissimus*
poisoning, there is a lack of coordination, vision trouble, trembling,
listlessness, abortion, inability to eat or drink, and paralysis of the
legs. (See also *Oxytropis.*)

Astragalus racemosus flower

Astragalus racemosus plant

47

Astragalus racemosus
seed
Astragalus racemosus
rootstock

Astragalus racemosus fruit

Astragalus bisulcatus plant

Astragalus bisulcatus fruit
Astragalus bisulcatus fruit

Astragalus bisulcatus
flower

48

Astragalus pectinatus
leaves, flowers

Astragalus pectinatus flower

Astragalus pectinatus fruit

Astragalus pectinatus
seed
Astragalus pectinatus flower

49

Astragalus mollissimus plant

Astragalus mollissimus
flower

Astragalus mollissimus
fruit

Astragalus mollissimus
seed

Crotalaria sagittalis L.
Rattlebox
FABACEAE

Herbaceous annual 1–3 dm high, simple or branched, hairy with spreading hairs. Leaves alternate, simple, lanceolate to lance-elliptic, 3–7 cm long, 10–15 mm wide, entire margin; stipules at the base of the leaf conspicuous, pointed, decurrent along the stem. Flowers axillary, 2–3 on a short stalk, leguminous, about 8 mm long, yellow; calyx lobes lanceolate, longer than the corolla; June–September. Fruit a leguminous, inflated pod 15–25 mm long, 1 cm thick, black at maturity, seeds detach and rattle in the pod; fruit stalk slender, recurved. Seeds obliquely heart-shaped, 2–2.5 mm long, 2 mm wide, dark brown, smooth, glossy.

Unused areas, sloping roadside banks, open spots in blackjack oak woods, sandy or gravelly soils. New England to South Dakota, south to Texas and Florida.

Other species: This is the only species of *Crotalaria* found in the central states. In the southern states there is *C. spectabilis* Roth which is a great deal more toxic than *C. sagittalis.* Along the East Coast are *C. mucronata* Desv. and *C. retusa* L. which have been proven toxic.

POISONING: The poisonous principle in this *Crotalaria* is unknown, but apparently all parts of the plant contain some type of toxin. It is possibly an alkaloid, monocrotaline, the same as in *C. spectabilis.* The seeds contain the greatest amount of the toxin and are dangerous in the pasture, in hay, or in mash made from contaminated grain. All classes of livestock are susceptible, including fowl.

SYMPTOMS: Excessive salivation, bloody feces, emaciation, low blood pressure and heartbeat, nasal discharge, and stupor. Death may occur within 2 or 3 days if a quantity of the material has been ingested.

Crotalaria sagittalis plant

Crotalaria sagittalis plant

Crotalaria sagittalis flower, fruit

51

Crotalaria sagittalis seed

Crotalaria sagittalis flower

Gymnocladus dioica (L.) K. Koch

Kentucky coffee tree

FABACEAE

A tree to 18 m, usually about 10 m when grown in the open. Branches spreading, coarse; twigs red-brown with large leaf scars; trunk either tightly ridged or with loose scales. Leaves alternate, bipinnately compound, 5–8 dm long, the basal leaflets often not divided; ultimate leaflets ovate, 4–5 cm long, 3–4 cm wide; young leaves densely hairy; leaf stalk greatly enlarged at the base. Flowers in a terminal panicle, 25–30 cm long, on new growth, inconspicuous; each flower about 1 cm across, whitish, the base tubular about 8 mm long; May. Fruit a leguminous pod 10–15 cm long, 4–5 cm wide, thick-walled, 1–8 seeds in a gelatinous substance which dries at maturity. Seeds circular, 15–20 mm across, 10–12 mm thick, dark olive-brown.

Rich soils of creek banks or rocky, open hillsides; often planted. New York to South Dakota south to Texas and West Virginia.

POISONING: The poisonous principle is unknown but has been reported to be cystine, an alkaloid. The leaves, especially of young sprouts, and the seeds with the gelatinous material around them contain the poison. Only a few cases of human poisoning have been reported from eating the seeds or using them as a coffee substitute. Farm animals have been poisoned from eating the young sprouts.

SYMPTOMS: In humans and livestock, gastric disturbances and pains, diarrhea, vomiting, irregular pulse, and a narcotic-like effect on the nervous system.

Gymnocladus dioica twig *Gymnocladus dioica* leaf

Gymnocladus dioica flower

Gymnocladus dioica fruit

Gymnocladus dioica
flower
Gymnocladus dioica seed

53

Lathyrus polymorphus Nutt.
FABACEAE
ssp. *incanus* (Sm. & Rydb.) C. L. Hitchc.
(L. *incanus* [Sm. & Rydb.] Rydb.)
Hoary peavine, wild pea

Herbaceous perennial, 15-30 cm tall, erect, not vining, hairy on all parts; stem not winged. Leaves alternate, pinnately compound, 3-5 cm long, 3-5 pairs of linear leaflets, 1.5-2 cm long, 2-3 mm wide, the tendril at the end of the leaf short and bristle-like, not twining. Flowers in a raceme of 2-5 flowers on a peduncle 5-7 cm long; flowers leguminous, 2-3 cm long, usually blue with some white on the petals; June–July. Pods 3-5 cm long, 8-10 mm wide, 4-7 seeds. Seeds globose, 5-6 mm diameter, dark brown, smooth, dull.

Plains or open woods, especially in sandy soil. South Dakota and Wyoming, south through Nebraska, Colorado, Kansas, Oklahoma, and Texas.

Other species: *L. ochroleucus* Hook., yellow vetchling. Pale yellow flowers, wide leaflets; found in southern Canada and the northern United States. *L. odoratus* L., garden sweet pea. Climbs by tendrils and has winged stem; seldom escapes or persists except for the variant *L. latifolius* L., perennial sweet pea. *L. palustris* L., marsh vetchling, and *L. venosus* Muhl., bushy vetch; both with purple flowers and found in the eastern part of the central states and northeast to New England. *L. pusillus* Ell., low peavine. Annual with purple flowers, winged stems, erect or sprawling, found in the Gulf states and as far north as Kansas and Missouri. These are the usual species listed as having been proven to contain poison.

POISONING AND SYMPTOMS: The toxic principle and symptoms of poisoning vary according to the species, but in all cases it is the seed that causes illness. In *L. polymorphus* the toxin has not been described, but horses have developed lameness from eating the plant with seeds. In *L. odoratus* and L. *pusillus* the poison is beta (gamma-L-glutamyl)-aminopropionitrile. This causes lameness, retardation of growth, paralysis, and skeletal deformities in livestock. In *L. latifolius* the poison principle is L-alpha, gamma-diaminobutyric acid and beta-aminopropionitrile. In farm animals, this combination causes hyperexcitability, convulsions, and death.

Lathyrus polymorphus
fruit

54

Lathyrus polymorphus
seed

Lathyrus polymorphus
plant

Lathyrus polymorphus leaf

Lathyrus ochroleucus flower

Lathyrus ochroleucus
seed
Lathyrus ochroleucus
leaf

55

Lathyrus odoratus leaves, stem

Lathyrus odoratus leaves

Lathyrus odoratus fruit
Lathyrus odoratus flower

Lupinus argenteus Pursh
Lupine, bluebonnet

FABACEAE

Herbaceous perennial from a stout rootstock, several erect stems from one base, 3–7 dm high; stems simple, appressed hairy. Leaves alternate, palmately divided into 6–10 leaflets; leaflets oblanceolate, 2–7 cm long, smooth or hairy above, densely hairy beneath. Flowers in racemes 10–25 cm long, terminal, loose or compact; flowers leguminous, 8–12 mm long, blue, purple or bicolored; flower stalk 4–8 mm long, hairy; June–July. Pods 2–3 cm long, 8–10 mm wide, densely hairy, indented between the seeds. Seeds flattened, irregularly circular or quadrangular, 3–5 mm across, 2–2.25 mm thick, cream-colored.

Dry prairies, unused areas, slopes in open prairie-woods. North Dakota and Montana, south to western Nebraska, eastern Colorado,

56

New Mexico, Arizona, the western tip of the Oklahoma panhandle, and the northwest portion of the Texas panhandle.

Other species: *L. pusillus* Pursh, dwarf lupine. An annual with only one stem from the ground, branched near the base; 10–20 cm tall. Open sandy areas. Washington to Saskatchewan, south to western Kansas, New Mexico, and northern Arizona. It has been reported as poisoning livestock in California.

POISONING: Lupine poisoning comes from a variety of alkaloids such as lupinine and lupanine, along with others. They are found in all parts of the plant, especially the seeds. The toxin is not destroyed by drying, and hay cut from prairies containing lupine is especially dangerous if mowed just before the seeds mature. Such prairies should not be pastured.

SYMPTOMS: Livestock experience labored breathing, nervousness, and convulsions. The symptoms may develop soon after ingestion, or they may not develop for several days. The animal may die quietly within 24 hours or thrash around for some time before death.

Lupinus argenteus plant

Lupinus argenteus leaf

Lupinus argenteus flowers

Lupinus argenteus flowers

Lupinus argenteus seeds

Lupinus argenteus fruits

Lupinus pusillus plant

Lupinus pusillus seeds

Melilotus officinalis (L.) Lam.   FABACEAE
Yellow sweet clover

Herbaceous biennial 1–1.5 m high, stems erect, branched, glabrous. Leaves compound with 3 leaflets, each obovate, 1–2.5 cm long, finely toothed, terminal leaflet on a stalk; petiole 1–2 cm long. Flowers in axillary racemes 5–12 cm long, usually arched; flowers leguminous, 5–7 mm long, yellow; June–August. Pods 2.5–3.5 mm long, veiny. Seeds oval, 2–2.25 mm long, 1–1.25 mm wide, yellow, smooth, the scar near one end.

Roadsides, fence rows, fields, and unused areas; also planted as a crop. Native of Europe, now established over most of southern Canada and all of the United States.

Other species: *M. alba* Desr., white sweet clover. Similar but coarser and with white flowers.

POISONING: Coumarin is produced in all parts of the plant and is a harmless substance. But due to frost, dry weather, or improper drying of hay, the plant may mold and dicoumarin is produced. Dicoumarin is an anticoagulent, and livestock that have ingested the clover may bleed to death internally or from being dehorned while infected.

SYMPTOMS: Internal bleeding, pockets containing blood swell under the skin, strong heartbeat with normal or subnormal temperature.

58

Melilotus officinalis leaf

Melilotus officinalis plant

Melilotus officinalis
flower

Melilotus officinalis fruit

Melilotus officinalis fruit

Melilotus officinalis
seed

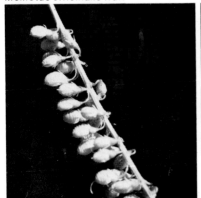

Oxytropis lambertii Pursh
Locoweed

FABACEAE

Stemless, herbaceous perennial, leaves and flower stems directly from the branched rootstock and strong taproot; 5-20 flower stems, each 2-3 dm high, equaling or overtopping the leaves. Leaves pinnate, 1-3 dm long, erect; 7-23 linear-lanceolate leaflets 1-4 cm long, 1-3 mm wide, appressed hairy. Flowering spikes 7-15 cm long, compact or loose; each flower leguminous, 18-22 mm long, sessile, blue, purple, or pale lavender; May-June. Pods cylindric, 2-3 cm long, usually pointed, erect, hairy, many seeds. Seeds irregularly kidney-shaped, 2-3 mm long and about as wide, dark brown, dull, smooth.

Prairies, roadsides, and open prairie-woods in dry situations. Minnesota, Manitoba, and Montana south to Arizona and Texas.

Other species: *O. sericea* Nutt., white locoweed. Similar but with silky hairy leaves and white flowers. British Columbia to Saskatchewan south to western Kansas and New Mexico.

POISONING: The toxic principle has not been clearly defined. Feeding experiments are the source of most of the positive information about poisoning. It appears to be cumulative, for the animal must eat a large amount of the plant over a long period of time. Horses are more susceptible than cattle or sheep and will develop a craving for the plant.

SYMPTOMS: Lack of coordination, trembling, vision trouble, inability to eat or drink, paralysis, listlessness, lack of interest unless startled—and then the animal becomes unruly. During these activities the back legs often buckle and the horse goes down on its haunches or even becomes completely prostrate. Death usually comes about a month after the first ingestion of the material.

Oxytropis lambertii
flower

Oxytropis lambertii plant

Oxytropis lambertii fruit

Oxytropis lambertii
rootstock

Oxytropis sericea leaves

Oxytropis sericea flowers

Oxytropis sericea seeds

61

Robinia pseudoacacia L.
Black locust

FABACEAE

Tree to 15 m high, with high, slender branches when crowded and with short, stout thorns at the leaf base. Leaves alternate, pinnately compound, 10–18 cm long, 9–19 leaflets; leaflets elliptic to oval, 2–3.5 cm long, entire margin. Flowers in drooping racemes 10–15 cm long, 10–30 flowers, leguminous, 1–2 cm long, white, sweet-scented; flower stalks 5–6 mm long; May–June. Pods 5–10 cm long, 1–1.5 cm wide, flat, thin-walled, a small point at the outer end; 4–9 seeds. Seeds kidney-shaped, 5 mm long, 3 mm wide, brown mottled with darker brown, smooth.

Well-drained soils, rocky hillsides, open areas, often planted as an ornamental or for fence posts. New York to South Dakota, south to Nebraska and Oklahoma, east to Texas and North Carolina.

POISONING: The plant contains the phytotoxin robin, the glycoside robitin, and the alkaloid robinine. The poisonous parts are the inner bark, young sprouts, wilted leaves, and seeds. Children have been poisoned from eating the seeds or chewing the inner bark as they might do with slippery elm. Livestock poisoning comes from stripping and eating the bark, eating young sprouts or wilted leaves, or eating the pods with the seeds.

SYMPTOMS: Dullness, vomiting, diarrhea with possible blood in the feces, labored breathing, weak pulse, coldness of extremities, and dilation of the pupils. Symptoms in humans and livestock are similar and appear shortly after eating the poisonous parts.

Robinia pseudoacacia
trunk

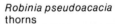

Robinia pseudoacacia
thorns

Robinia pseudoacacia flowers

Robinia pseudoacacia flowers

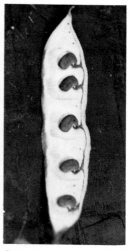

Robinia pseudoacacia flowers

Robinia pseudoacacia fruits

Robinia pseudoacacia
opened fruit

Robinia pseudoacacia
seeds

63

Phoradendron tomentosum (DC.) Gray LORANTHACEAE
Mistletoe

Woody perennial, semiparasitic on several species of trees; branches green, granular. Roots embedded in the branches of the host tree. Leaves opposite, oval, 3–4.5 cm long, 1.5–2.5 cm wide, fleshy, granular, evergreen. Flower spikes axillary, jointed, 2–3 cm long; flowers embedded in the spike, yellow-green, inconspicuous; March–July. Fruits globose, 4–6 mm diameter, white, translucent, sticky when crushed, single seed. Seed flattened ovoid, 2.5–3 mm long, dark green with a white papery covering.

On branches of trees such as elm, oak, cottonwood, hackberry, ash, and sycamore, either upland or along streams. Mexico to Texas, north and east to southeast Kansas, Missouri, Arkansas, and Louisiana.

Other species: There are several other species in other parts of the United States, all of which may be used for Christmas decoration. All are similar and not easily mistaken for other plants. Their combined range is from New Jersey across the southern United States to California and Mexico.

POISONING: The plants contain the amines beta-phenylethylamine and tyramine. Poisoning, and at least one known human death, has been reported from eating the berries or making them into a drink.

SYMPTOMS: Gastroenteritis, diarrhea, and weak pulse.

Phoradendron tomentosum leaves, flowers

Phoradendron tomentosum flowers

64

Phoradendron tomentosum fruits

Phoradendron tomentosum seeds

Euphorbia spp.
Spurges

EUPHORBIACEAE

There are many species of spurges in the central states and they are so variable that a description of one would hardly help in identifying another. The *Euphorbia* flower is distinctive and in most cases the "flower" is a cluster of tiny flowers in a cup-like structure. Some species have showy petaloid appendages on the margin of the cup. This gives the impression that the cluster is a single flower.

All spurges native to the central states are herbaceous and may be either annual or perennial. The leaves are all simple and both leaves and stem contain a milky juice. Our plants may be divided into two groups, the erect plants and the prostrate, or mat, spurges. The poisoning and symptoms are practically the same for all species.

E. corollata L., flowering spurge. Perennial, erect, 3–6 dm high, branched only at the top, leaves elliptic, 3–5 cm long, flower appendages conspicuous, white. Seeds ovoid, 2.5–3 mm long, mottled gray-brown. Prairies and roadsides. New York to Minnesota, south through Kansas to east Texas and Florida.

E. cyathophora Murr., painted leaf. Annual, erect or ascending, 5–8 dm tall; stem leaves linear to lanceolate, 4–12 cm long, uppermost leaves are panduriform, typically red at the base as is the poinsettia. Seeds ovoid, 3 mm long, brown, tuberculate. Moist or shaded soil. Wisconsin, South Dakota, south to Texas, Florida, Virginia, and Indiana. (*E. heterophylla* L.)

E. cyparissias L., cypress spurge. Perennial and colonial from rhizomes; erect stems 1–3 dm high, branched at the top, densely leafy; stem leaves linear, 1–3 cm long; floral leaves and bracts cordate, 4–10 mm broad, broader than long, often reddish. Seeds oblong-cylindric, about 2 mm long, gray-brown, smooth. Roadsides and unused ground. Maine to North Dakota, south to Colorado, and east to Missouri and Virginia.

65

E. dentata Michx., toothed spurge. Annual, erect or ascending, 2–5 dm high, branched from the base on up. Leaves opposite, linear-lanceolate to rhombic-ovate, 1–9 cm long, coarsely dentate. Seeds broadly ovoid, 2–2.5 mm long, black, heavily tuberculate and wrinkled. Roadsides, fields, and farmsteads. New York to South Dakota and Wyoming, south to Mexico, east to Louisiana and Virginia.

E. maculata L., mat spurge, wartweed. Annual, prostrate or slightly ascending, the hairy stems often 4 dm long, forming mats. Leaves elliptic-ovate, 4–15 mm long, 2–6 mm wide, often a red spot near the center, minutely toothed; ovary hairy. Seeds about 1 mm long, angled and transversely ridged, light silvery brown. Roadsides, fields, unused ground, farmsteads. Quebec to North Dakota, south to Texas and Florida. (*E. supina* Raf.)

E. marginata Pursh, snow-on-the-mountain. Annual, erect, 3–9 dm high, branched above the middle. Stem leaves ovate to broadly elliptic; leaves and bracts directly beneath the flowers whorled, with showy, wide, white margins. Flower appendages white, conspicuous. Seeds broadly ovoid, 3.5–4 mm long, light brown, tuberculate. Sandy or rocky soils; fields, prairies, roadsides, and farm lots. Minnesota to Montana, south to New Mexico, Texas, and Missouri.

E. milii Des Moul., crown of thorns. Cultivated as a potted plant indoors in the central states. Stem thorny. Leaves on young growth only, broadly elliptic, 1–3 cm long. Flowers conspicuous because of the large, salmon-red appendages. Not a serious threat in poisoning.

E. nutans Lag., eyebane. Annual, mainly erect, 1–5 dm tall, branched. Leaves opposite, oblong, 5–35 mm long, often curved, base unequal, usually a red spot near the middle, toothed. Flowers axillary, inconspicuous. Seeds ovoid, 1–1.25 mm long, angles rounded, faces with low, irregular, transverse ridges, dark brown to black. (*E. preslii* Guss.) Because of a mixup in nomenclature, *E. nutans* is often given as *E. maculata* which is a mat spurge. Waste, weedy areas especially around farmsteads. Common in the eastern United States.

E. podperae Croiz., leafy spurge. Perennial, erect, 3–7 dm high, branched. Stem leaves broadly linear, 3–7 cm long, 2–9 mm wide; leaves below the umbel ovate, 1–2 cm long, 5–10 mm wide; bracts beneath the flowers broadly cordate, 10–15 mm long and as wide. Seeds ovoid, 2–3 mm long, 1.5 mm wide, smooth, brown or silvery brown and mottled. Weedy prairies, roadsides, railroad embankments. Across the northern United States from New England to California and south to Colorado, Kansas, and Maryland. (*E. esula* L.)

E. prostrata Ait., mat spurge. Annual, prostrate forming mats 3–5 dm across, stems hairy. Leaves opposite, oblong, 4–12 mm long, 3–7 mm wide, minutely toothed. Capsule hairy at the base and on the angles. Seeds ovate, about 1 mm long, sharply angled, faces with irregular, sharp, transverse ridges. Roadsides, pastures, farmsteads, and unused city lots. Scattered in central states from Nebraska to Texas and Louisiana.

E. pulcherrima Willd., poinsettia. Cultivated as a potted plant in the central states, outdoors in warm climates. Leaves generally ovate, lobed, those beneath the flowers partially or entirely bright red or occasionally white. Only one human death has been reported from eating the leaves of the poinsettia.

POISONING: The toxin is an acrid principle, apparently not yet fully defined. It is not destroyed by drying. All parts of the plant are poisonous to some extent; the milky juice may cause dermatitis with blisters; also involved in photosensitization in lambs.

SYMPTOMS: In humans, severe irritation in the mouth, throat, and stomach, also vomiting. Deaths are extremely rare. In livestock, vomiting, diarrhea, pain in the digestive tract, scours, emaciation, weakness, and possible loss of hair around the feet.

Euphorbia corollata flowers

Euphorbia dentata seeds

Euphorbia dentata plant   *Euphorbia dentata* fruits   67

Euphorbia maculata
seeds

Euphorbia maculata
plant

Euphorbia marginata plant   *Euphorbia marginata* fruits

Euphorbia marginata   *Euphorbia milii* stem,   *Euphorbia milii* flowers
seeds   leaves

68

Euphorbia nutans seeds

Euphorbia nutans plant

Euphorbia podperae plant

Euphorbia podperae flowers

Euphorbia prostrata leaves, flowers

Euphorbia pulcherrima leaves

Euphorbia prostrata seeds

69

Ricinis communis L.
Castor bean

EUPHORBIACEAE

Herbaceous annual with one central stem, 2-4 m high, stem hollow, purplish with a thin waxy covering easily rubbed off, a ring around the stem at the leaf attachment. Leaves alternate, circular in outline, 30-65 cm across, usually with 9 pointed lobes, a main vein to each, margin toothed; leaf stalk 3-5 dm long, purplish, attached to the blade off center, the veins radiating from there to the lobes. Flowers in racemes 8-15 cm long at the end of the main stem; late summer. Fruit raceme erect, 15-20 cm long, each fruit oblong, 2-3 cm long, covered with short, fleshy spines; usually 3 seeds. Seeds oval, 6-10 mm long, 4-7 mm wide, varying in size and color according to varieties; light brown and mottled or streaked with dark brown, or the reverse; caruncle (point of attachment) obvious. Garden plant throughout the United States.

POISONING: The leaves and seeds contain the phytotoxin ricin. In its pure state it is extremely poisonous, and even in the castor bean, 2-4 beans can cause lethal poisoning in children. Children do play with the beans and are especially subject to poisoning. Farm animals are poisoned from eating either the leaves or seeds. If castor beans are grown around the home, the flowers should be removed and destroyed.

SYMPTOMS: Burning of the mouth and throat, thirst, vomiting, stomach pains, diarrhea, dullness, weakness, and a weak, rapid pulse occur in both humans and livestock.

Ricinis communis plant

Ricinis communis leaves

70

Ricinis communis fruits

Ricinis communis flowers

Ricinis communis seeds

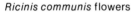

Parthenocissus quinquefolia (L.) Planch.   **VITACEAE**
Virginia creeper, five-leaved ivy

Woody vine clambering over fences or climbing into trees by means of tendrils and aerial roots to a height of 18–20 m. Aerial roots branched, a disk on the end of each branch. Leaves alternate, palmately compound, the 5 leaflets obovate, 8–13 cm long, 4–6 cm wide, coarsely toothed; leaf stalk 10–20 cm long. Flowers in a panicle opposite a leaf, panicle 6–12 cm long with a definite central axis and side branches; flowers 5–7 mm across, greenish or green-bronze; June–July. Fruit subglobose, 5–8 mm diameter, dark purple with a waxy bloom, 4 seeds. Seeds ovoid with 1–2 flat surfaces, 4–5 mm long, 3–4 mm wide, dark chocolate brown, an ovate marking on the rounded side.

Woods, fence rows, and old farmsteads in either rich or rocky soils. Maine to Minnesota, south to South Dakota, Nebraska, Texas, and Florida.

Other species: *P. vitacea* (Knerr) Hitchc., woodbine. Similar but the tendrils have no disks and the flower and fruit panicle has no central axis but continues to divide into 2 subequal branches.

POISONING: The fruits apparently contain an unidentified toxin. Experimental animals have been poisoned fatally when fed the berries. Reports of children being fatally poisoned are on record, but in all cases there is some question about the exact identification of the plant or there are other circumstances which cause doubt.

SYMPTOMS: None have been given in the known cases. 71

Parthenocissus
quinquefolia tendril

Parthenocissus
quinquefolia branch

Parthenocissus
quinquefolia plant

Parthenocissus
quinquefolia seed

Parthenocissus vitacea
flowers

Parthenocissus vitacea flowers

Parthenocissus vitacea fruits

Aesculus glabra Willd.   HIPPOCASTANACEAE
var. *arguta* (Buckl.) Robins.
Buckeye

Small tree or shrub 3–8 m high in our area, taller in the east. Branches tan, coarse, bark of young trees scaly. Leaves opposite, palmately compound, dark green and glossy; 5–7 leaflets, 7–15 mm long, toothed; leaf stalk as long as the leaflets. Flowers in a terminal panicle 10–12 cm long, often densely flowered; flowers bilaterally symmetrical, 20–25 mm long, yellow, the upper petals with an orange spot or streak; stamens and pistil extended beyond the petals; April–May. Fruit globular, 3–4.5 cm diameter, rusty brown, somewhat spiny, at maturity the thick wall splits into 3–4 sections. Seeds globose with 1–3 flattened surfaces, 2–2.5 cm diameter, hard, red-brown with a large, pale brown scar.

Stream banks, brushy hillsides, and ditches, in rich, rocky, or clay soils. Pennsylvania to Michigan, southwest to Nebraska and Oklahoma, and east to Mississippi, Tennessee, and Kentucky.

POISONING: The young sprouts, leaves, flowers, and seeds contain the glycoside esculin. People have been poisoned from eating the seeds or drinking tea made from the leaves. Livestock poisoning is reported from eating the young sprouts on recently cleared land or from eating the seeds. Honey made from the flowers of some species of *Aesculus* may also be poisonous.

SYMPTOMS: In humans and livestock, weakness, twitching of muscles, lack of coordination, depression, diarrhea, vomiting, inflammation of mucous membranes, dilated pupils, paralysis, and stupor.

Aesculus glabra trunk

Aesculus glabra flowers, leaves

73

Aesculus glabra fruits

Aesculus glabra flowers

Aesculus glabra seed

Toxicodendron radicans (L.) O. Ktze.   ANACARDIACEAE
ssp. negundo (Greene) Gillis
Poison ivy

Perennial woody vine up to 20 m long, or a shrub-like plant 0.5–2 m high; aerial roots along the climbing stem. Leaves alternate, compound with 3 leaflets; leaflets ovate, 12–17 cm long, 7–10 cm wide, short hairy, margin coarsely toothed or lobed, tip pointed; leaf stalk 6–20 cm long, hairy. Flowers in axillary panicles on new growth, panicles 5–10 cm long; flowers yellow-green, 6–8 mm across, petals recurved; May–June. Fruits globose, 4–5 mm diameter, gray or cream-white, outer covering papery with shallow, longitudinal grooves, inner coat fibrous.

Woods, fence rows, pastures, thickets, and unused areas. This and its varieties or similar species found throughout most of North America.

Other species: *T. rydbergii* (Small ex Rydb.) Greene. A subshrub with suborbicular or broadly ovate leaflets and glabrous leaf stalks. *T. toxicarium* (Salisb.) Gillis, poison oak. A subshrub with the teeth and lobes of the leaves rounded. Found from Missouri east and south through the southeastern states; occurs in the Plains states only in the southeast corner of Kansas.

POISONING: The poisonous principle is 3-n-pentadecylcatechol and is found in all parts of the plant. When the resin ducts are broken the material exudes and may come in contact with the skin. The oil is not volatile but tiny drops of it may be carried on the fur of animals, on clothing, on guns or other articles, and on particles of smoke

74

when the plant is burned. People vary in their susceptibility to the poison, and even that may vary from time to time. A person can contract dermatitis from the plant at any time of the year. The leaves must not be eaten to "build up immunity" because they often cause serious gastric disturbance.

SYMPTOMS: In humans, itching, burning, redness of skin, and small or large blisters. Secondary infection may result from scratching. Livestock are not susceptible to the poison.

Toxicodendron radicans winter plant
Toxicodendron radicans summer plant

Toxicodendron radicans leaves

Toxicodendron radicans aerial roots

75

Toxicodendron radicans
fruits

Toxicodendron radicans
seeds

Toxicodendron toxicarium leaves   *Toxicodendron toxicarium* fruits

---

Linum usitatissimum L.   LINACEAE
Common flax

Herbaceous, slender annual, 6-10 dm high from a slender taproot; main stem leafy, branched near the top, but often several branches come from near the base. Leaves alternate, lance-linear, 15-30 mm long, 2-4 mm wide; lower leaves shed before fruiting time. Flowers regular, 2-2.5 cm across, 5 light blue petals with dark blue lines at the base; July–October. Capsule globose, 6-10 mm diameter, calyx persistent, 7-9 mm long. Seeds narrowly ovate, flattened, 4-5 mm long, 2-2.5 mm wide, brown, smooth, semiglossy.

Commonly cultivated in the northern states for the fiber, linseed oil, or press cake made from the seed. May escape from fields or from gardens further south.

Other species: *L. perenne* L. var. *lewisii* (Pursh) Eat. & Wright, blue flax. Native perennial, erect or decumbent, blue flowers. Dry prairies or unused land. Wisconsin to Alaska, California, and Texas.

Loss of sheep has been attributed to this plant. *L. rigidum* Pursh, stiff-stemmed flax. Slender annual, 3–5 dm high, yellow flowers; some varieties are short and bush-like with dark centers in the flowers. Open prairies. Manitoba to Alberta, south to Colorado, New Mexico, and Texas, north through Missouri to Minnesota. Several other species grow in the central states but they have not been involved in poison studies or diseases.

POISONING: The poisonous principle is linamarin, a cyanogenetic glycoside. It is found mainly in the leaves and the chaff screenings from the production of linseed oil. The plant is potentially dangerous when used as hay. Press cake is a valuable feed and is made harmless by heat treatment; it should be purchased from a reliable source.

SYMPTOMS: Rapid and difficult breathing, gasping, staggering, paralysis, coma, and death.

Linum usitatissimum flower
Linum usitatissimum leaves

Linum usitatissimum seeds

Linum usitatissimum fruits

77

Linum perenne leaves

Linum perenne flowers

Linum perenne fruits

Linum perenne flowers

Linum perenne seeds

Linum rigidum seeds

Linum rigidum fruits

Cicuta maculata L.
Water hemlock

APIACEAE

Herbaceous perennial 8-20 dm high from a fascicle of 2-8 tuberous roots, tubers 3-8 cm long, 5-10 mm thick. Stem stout, glabrous, often with purple spots, hollow with partitions at the nodes; nodes close together in the swollen stem base. Leaves alternate, 1-3 times pinnately compound, 2-5 dm long; ultimate leaflets lance-ovate, 5-10 cm long, 1-2 cm wide, sharply toothed, glabrous; leaf stalk 5-7 cm long, the base sheathing the stem. Flowers in compound umbels at branch ends; umbels 5-20 cm across, round or flat-topped, becoming somewhat spherical in fruit; flowers 3-4 mm across, 5 white petals; July-August. Fruits ovoid or ellipsoid, flattened, several rounded ribs on the surfaces.

Wet meadows, banks of streams, wet ditches, and edges of ponds and lakes. Common in the central states; ranges from eastern Canada to Manitoba, south to Texas and Florida.

Other species: *C. bulbifera* L., bulbous water hemlock. Similar but with narrow leaflets and the production of bulblets in the leaf axils. Newfoundland to British Columbia, south to Oregon, Iowa, and Pennsylvania. *C. douglasii* (DC.) Coult. and Rose. Similar to *C. maculata* but the range is western, overlapping with *C. maculata* in the Dakotas.

POISONING: The poisonous principle is cicutoxin, a resinoid which affects the nervous system. It is found in all parts of the plant but is concentrated in the tuberous roots. Since the plant grows in wet places, these tubers are often pulled up when an animal eats the tops. It is reported that one good bite is sufficient to kill a human. The tubers resemble those of a Jerusalem artichoke, and people have occasionally eaten them. Children have also been poisoned by whistles made from the hollow stems. In extreme cases death may occur in less than an hour.

SYMPTOMS: In both humans and livestock, salivation, twitching and stiffening of muscles, diaphragm contractions, dilation of pupils, violent convulsions, slow breathing, and dizziness.

Cicuta maculata leaf

Cicuta maculata flowers

79

Cicuta maculata plant

Cicuta maculata
hollow stem

Cicuta maculata seeds

Cicuta maculata fruit head

Cicuta maculata tuberous roots

Conium maculatum L.
Poison hemlock

APIACEAE

Herbaceous biennial from a simple, stout taproot; stems 1.5–3 m tall, many branches, hollow except at the nodes, glabrous, purple spotted. Leaves alternate, broadly triangular in outline, 2–4 dm across, 3–4 times pinnately dissected, the ultimate segments 3–6 cm long, 1 cm wide, coarsely toothed; the leaves have a fern-like appearance. Flowers in compound, flat-topped umbels 4–6 cm across, at the ends of the branches; each flower 2–3 mm across, 5 white

80

petals; June–August. Fruits pale brown, broadly ovoid, 1.5–3 mm long, easily separated into 2 sector-shaped parts, ventral side smooth, dorsal side with 5 obvious, light colored ribs.

Waste areas, cow lots, and roadside ditches where the soil is often moist; grows in drier soil than *Cicuta.* Introduced from Europe and now widespread in the United States and southern Canada except in high mountains.

Conium is often confused with *Daucus carota* L., the wild carrot, but the latter is definitely a hairy plant and the early stages of the flowering umbel are cupped like a bird's nest.

POISONING: Five alkaloids are found in *Conium:* lamba-coniceine, coniine, N-methyl coniine, conhydrine, and pseudoconhydrine. All parts of the plant are poisonous. The roots contain the least amount. The stems and leaves are most poisonous just before the fruits mature. The geographic location appears to have an affect on the amount of poison: southern plants are considered more dangerous than the northern ones. This is probably the plant used to kill Socrates.

SYMPTOMS: In both humans and livestock, loss of appetite, salivation, gastrointestinal irritation, lack of coordination, respiratory difficulties, slow and feeble pulse, and a "mousy" odor of the urine. Death comes from paralysis of the lungs.

Conium maculatum
hollow stem

Conium maculatum plant

Conium maculatum leaf   81

Conium maculatum flower

Conium maculatum flower

Conium maculatum
seeds

Conium maculatum fruits

Conium maculatum fruits

Nerium oleander L.
Oleander

APOCYNACEAE

A potted shrub in the central states but grown outdoors in the extreme southern United States. May reach a height of 4 m, but usually is kept trimmed back. Leaves in whorls of 3, elliptic, 15-20 cm long, 3-4 cm wide; midrib prominent, lateral veins close and parallel; dark, glossy green above, yellow-green below. Flowers in clusters of 5-8, each flower 4-5 cm across, pink or white, occassionally streaked; any season for flowering, depending on the place of growth.

POISONING: All parts of the plant contain the cardiac glycosides oleandroside and nerioside. The plants are extremely poisonous, a single leaf may be lethal to a human. People have been poisoned from using the woody sticks for roasting weiners. All classes of livestock are susceptible and should be kept away from the bushes. In experimental work, sheep have been killed by feeding oleander leaves at the rate of 0.005 percent of the animal's weight. In Hawaii, *N. indicum* Mill. is most common, and both people and livestock have died from eating the leaves.

SYMPTOMS: In humans and livestock, depression, vomiting, gastroenteritis, increased pulse rate, dizziness, bloody feces, and paralysis of the respiratory muscles. Death usually occurs within one day.

Nerium oleander plant

Nerium oleander leaves

83

Nerium oleander leaves

Nerium oleander flowers

Nerium oleander flowers

ASCLEPIADACEAE

Asclepias spp.
Milkweeds

There are many species of milkweeds in the central states and all should be considered potentially dangerous. Some of them have been known to cause sickness or death and others have been analyzed and found to contain poisons. The various species do not necessarily resemble each other but they all contain a milky juice, the seeds have a tuft of silky hairs, the flowers are in umbels, and each flower has the calyx and corolla reflexed or spreading. Arising from the corolla is a crown (hood) which contains the stamens and pistil as well as other structures. The genus could be divided into the broad-leaved species and the narrow-leaved species, the latter having linear leaves.

A. asperula (Dcne.) Woods. Herbaceous perennial, decumbent, broad leaves; flowers terminal, corolla pale yellow-green; hoods purple, spreading, clavate. Dry soils from southwest Kansas on south and west. Definitely poisonous.

A. engelmanniana Woods. Herbaceous perennial, erect, 8–12 dm high, leaves linear, 12–18 cm long; flowers axillary, corolla lobes about 5 mm long, pale green; pods slender. Dry soils from western South Dakota south and west to Arizona and Mexico. Suspected of being poisonous.

A. incarnata L., swamp milkweed. Herbaceous perennial, up to 2 m high, slender, branched; flowers terminal on branches, corolla lobes 3–5 mm long, pink. Swamps and wet stream banks ranging from the eastern edge of the Rocky Mountains to the East Coast. Suspected of causing death in sheep.

A. latifolia (Torr.) Raf., broadleaf milkweed. Herbaceous perennial, erect to 6 dm, stout; leaves nearly as wide as long, indented at the end; flowers axillary, corolla lobes 4–5 mm long, pale green to nearly white; pods stout. Dry, sandy soils from Nebraska to Texas, west to Utah and California. Definitely poisonous.

A. pumila (Gray) Vail, dwarf milkweed. Herbaceous perennial from deep rootstocks; plant 1–3 dm high, leaves linear, 2–4 cm long, crowded on the stem. Flowers axillary at the uppermost nodes, corolla lobes 2–3 mm long, white or tinged with pink; pods slender. Dry soils from North Dakota and Wyoming to Texas and New Mexico. Definitely poisonous.

A. speciosa Torr., showy milkweed. Herbaceous perennial, erect, about 1 m high, broad leaves. Flowers axillary, corolla lobes 1–1.5 cm long, pink, hoods erect or ascending, pointed, 1–1.3 cm long, obvious; pods stout. Dry or moist soils from Minnesota to British Columbia and south to California and Texas. Experiments have shown it to be poisonous.

A. stenophylla Gray, narrow-leaved milkweed. Similar to *A. engelmanniana* only smaller in all respects; leaves 6–14 cm long. Suspected of being poisonous.

A. subverticillata (Gray) Vail, poison milkweed. Herbaceous perennial from a stout rootstock, stems to 1 m high, leafy throughout; leaves in whorls of 3–5, linear, 2–13 cm long, short vegetative branches in some of the axils; flowers axillary, corolla lobes 3–5 mm long, white; pods slender, long tapered. Dry sandy soils from western Kansas to Utah, Arizona, and Texas. Definitely poisonous.

A. sullivantii Engelm., smooth milkweed. Herbaceous perennial, erect, 1–1.5 m tall; leaves broad, glabrous; flowers terminal and axillary, corolla lobes 8–10 mm long, pink; pods stout, smooth. Dry prairies from Ontario to North Dakota, south to Oklahoma, east to Missouri and Ohio. Reported as poisonous.

A. syriaca L., common milkweed. Similar to *A. sullivantii* but leaves densely hairy beneath and the pods rough with soft processes. Dry prairies, New Brunswick to North Dakota, south to Oklahoma, and east to Georgia and Virginia. Only circumstantial evidence of poison.

A. tuberosa L., butterfly milkweed. Herbaceous perennial with 85 several stems from one base; stem and leaves densely hairy; grows to 5 dm high; leaves lanceolate; flowers terminal, corolla lobes 7-8 mm long, yellow to orange-red; pods smooth, fusiform. Dry areas, east of a line from North Dakota to Arizona. Usually considered poisonous.

A. verticillata L., whorled milkweed. Similar to *A. subverticillata* but lacks the axillary vegetative shoots; up to 8 dm high, leaves 1.5-7 cm long. Dry or moist soil, across southern Canada to North Dakota, south to Texas, and east to Florida. Definitely poisonous.

A. viridiflora Raf., green-flowered milkweed. Herbaceous, erect perennial to 8 dm high; leaves broad, oval; flowers axillary, corolla lobes 6-7 mm long, sharply reflexed, yellow-green, often tinged with brown. Dry prairies, eastern Canada to Montana, south to Arizona and Georgia. Reported as poisonous.

A. viridis Walt., antelope horn. Herbaceous perennial from a stout rootstock, decumbent to ascending, stems to 5 dm long, several from one base; flowers similar to those of *A. asperula* but the hoods are not clavate and are more ascending from their base. Dry prairies, Tennessee and Florida to east Texas, northwest to Kansas and Nebraska. Suspected of being poisonous.

POISONING: Resinoids, glycosides, and a small amount of the alkaloids are present in all parts of the plant. They are not completely destroyed by drying although the toxicity may be reduced. The plants are quite distasteful to humans so there is little chance of people being poisoned. The plants grow in prairies used for pastures or mowed for hay and are, therefore, available to farm animals of all classes, even turkeys.

SYMPTOMS: Dullness, weakness, bloating, inability to stand or walk, high body temperature, rapid and weak pulse, difficult breathing, dilated pupils, spasms, and coma.

Asclepias asperula flower

Asclepias asperula plant

Asclepias incarnata fruit

Asclepias latifolia plant

Asclepias latifolia leaf

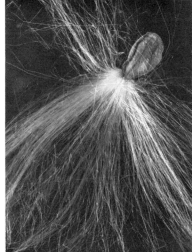

Asclepias pumila plant

Asclepias latifolia seed

Asclepias pumila seed

Asclepias speciosa flower

87

Asclepias subverticillata stem, leaves

Asclepias subverticillata flowers

Asclepias sullivantii flower

Asclepias sullivantii leaf

Asclepias tuberosa flower

Asclepias tuberosa fruit

Asclepias viridis fruit

Asclepias verticillata leaves

Asclepias viridiflora flowers

Datura stramonium L.
Jimsonweed

SOLANACEAE

Herbaceous annual, branched, the top spreading like a tree, 8–12 dm high, coarse, glabrous. Leaves alternate, ovate with irregular, pointed lobes or teeth, 8–20 cm long, 6–12 cm wide. Flowers single, axillary, erect or ascending on a short stalk; corolla funnel-shaped, 6–8 cm long, the flared end 4–5 cm across, 5-angled, often with narrow points, white or tinged with purple; July–October. Fruit ovoid, 4–5 cm long, 25–35 mm thick, erect, spiny, walls hard when mature, splits from the top into 4 sections exposing the many seeds. Seeds circular with one edge flattened, 3–3.5 mm across, 1 mm thick, black, the surface slightly rugose and minutely pitted.

Overgrazed pastures, feed lots, unused areas. New England across southern Canada to North Dakota, south to Texas and Florida, also on the West Coast. Other species of Datura not found in the central states are equally poisonous.

POISONING: All parts of the plant contain alkaloids among which are hyoscyamine, hyoscine (scopalamine), and atropine. As little as 4–5 grams of leaf or seed is sufficient to kill a child. Sucking nectar from the flowers may cause serious illness. Farm animals seldom eat the plant, but since one of its main habitats is the farm lot, the plant is readily accessible to them.

SYMPTOMS: Excessive thirst, vomiting, dilated pupils, dizziness, incoherence, hallucinations, rapid and weak pulse, inability to urinate, and eventual convulsions. These same general symptoms apply to humans, livestock, and poultry.

Datura stramonium plant

Datura stramonium
flower

Datura stramonium leaf

Datura stramonium flower

Datura stramonium fruit

Datura stramonium fruit

Datura stramonium seed

Hyoscyamus niger L.   SOLANACEAE
Henbane

Herbaceous annual or biennial, 8–10 dm high, several coarse branches from one base; stems hairy and often sticky. Leaves alternate, simple, sessile, 6–20 cm long, 3–14 cm wide, irregularly toothed or lobed with triangular lobes; leaves smaller on the upper stem. Flowers bell-shaped, 3 cm long, 2 cm across at the top, yellow-green with purplish veins and dark center; axillary along the upper 5–6 dm of the stem; June–July. Calyx enlarges and becomes urn-shaped in fruit, 2–3 cm long, 9–12 mm diameter, 5 points on the margin; capsule globose, 7–10 mm across, surrounded by the calyx, top of capsule breaks off as a cap to release the seeds. Seeds irregularly circular, flattened, 1.25–1.5 mm across, 0.5 mm thick, gray-brown, surface roughened with minute ridges.

Dry roadsides, prairies, and unused places, especially in slightly sandy soil. Native of Europe, now spread in scattered areas across southern Canada and the northern United States; fairly common in North Dakota and western South Dakota.

POISONING: All parts of the plant, including seeds, contain hyoscyamine, hyoscine, and atropine. People and all classes of farm animals have been poisoned from eating the leaves or seeds.

SYMPTOMS: Headache, nausea, increased pulse rate, convulsions, and coma precede death.

Hyoscyamus niger
flowers

Hyoscyamus niger leaves

Hyoscyamus niger
flowers

91

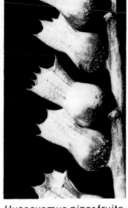

Hyoscyamus niger fruits   Hyoscyamus niger seeds

Hyoscyamus niger fruits

Physalis spp.   SOLANACEAE
Ground cherries

There are numerous species and varieties of the ground cherry in the central states as well as throughout the United States. Most of them are similar enough that they can be recognized as belonging to the genus. The general characteristics of those discussed here are: drooping, bell-shaped flower; calyx enlarging in fruit to form a bladder-like covering over the fruit; the fruit is berry-like with many seeds; all species described here are perennials.

P. heterophylla Nees. Herbaceous from an underground rhizome, stems 2–5 dm high, erect, branched, densely hairy. Leaves alternate, ovate, 3–8 cm long, broad teeth with shallow sinuses. Flowers axillary, 15–20 mm across, usually 5 small points on the margin, yellow with a dark center; June–September. Fruit stalk 10–15 mm long, elongating to 2–3 cm in fruit. Enlarged calyx 3–4 cm long, 2–2.5 cm thick; berry yellow, globose, 5–8 mm diameter, smooth. Seeds flattened, irregularly oval, about 2 mm long, 1.25 mm wide, yellow, minutely pitted and roughened. Dry soil, open woods, prairies, roadsides, and waste ground. New England to Saskatchewan, south to Colorado, Texas, and South Carolina.

P. hederaefolia Gray. Similar to the above but smaller. Short grass prairies in sandy soil from Montana and South Dakota south to New Mexico and Texas.

P. virginiana Mill. (*P. longifolia* Nutt.) A taller plant, 3–6 dm high, glabrous or with short hairs, branches high, resembling a miniature tree; leaves tapering at the base. Fields, unused areas, and fence rows; New York and Ontario to eastern Montana and south to Texas and Florida.

P. lobata Torr. *(Quincula lobata* [Torr.] Raf.). A decumbent plant with purple flowers. Sandy roadsides and broken areas in prairies.

92

Kansas and Colorado south to Texas, Arizona, and Mexico. Not usually considered dangerous.

POISONING: All parts of the plant contain the glyco-alkaloid solanine. The unripe fruit is the most dangerous, and children have been poisoned from eating it. The mature fruits of some species are edible. Most farm animals avoid the plant, but cattle have become sick after eating a plant with immature fruits.

SYMPTOMS: In livestock and humans, stomach and intestinal irritation and inflammation, troubled breathing, trembling, weakness, and paralysis.

Physalis hederaefolia
plant

Physalis hederaefolia
leaf

Physalis hederaefolia flower

Physalis hederaefolia fruit

Physalis hederaefolia
seed

93

Physalis heterophylla leaf

Physalis heterophylla flower

Physalis lobata plant

Physalis heterophylla flower

Physalis lobata seed

Physalis virginiana var.
sonorae plant

94

Physalis virginiana var.
sonorae flower, fruit

Physalis virginiana var. hispida
flower

Physalis virginiana var. hispida fruit

Solanum spp.
Nightshades

SOLANACEAE

The genus *Solanum* contains a number of poisonous species, the poisonous principle being approximately the same in all of them. For that reason, a few species are described briefly, then the toxin and symptoms are listed for the group.

S. carolinense L., horse nettle. Herbaceous perennial 5-8 dm high, from a deep taproot and rhizome 2-3 dm below ground. Stem spiny and more or less hairy. Leaves alternate, ovate, 7-16 cm long, 5-8 cm wide, margins sinuate or with shallow lobes; main veins with prickles. Flowers in short racemes near the top of the plant; flowers rotate, 15-25 mm across, petals united, 5 points on the margin, pale purple or white; flower stalks 10-15 mm long, usually nodding; June-August. Fruit a globose berry 10-15 mm diameter, yellow when mature, often persisting throughout the winter, many seeds. Seeds irregularly circular, 2-2.5 mm across, flattened, yellow or brownish, smooth or minutely pitted. Fields, roadsides, unused areas. Ontario to South Dakota, south to Texas and Florida.

S. americanum Mill. *(S. nigrum* L.), black nightshade. Glabrous annual, much branched, erect to 4 dm or trailing to 2 m, spineless; leaves ovate, sinuate, 5-9 cm long; flowers white, 4-6 mm across; 95

fruits dark purple, 7–9 mm diameter. Shaded, moist soil. Most of the United States east of the Rocky Mountains and on the Pacific coast.

S. elaeagnifolium Cav., silverleaf nightshade. Similar to *S. carolinense* but shorter, has narrower leaves, the whole plant white hairy, the flowers purple. Prairies, roadsides, and unused areas. Missouri and Kansas to the Southwest.

S. rostratum Dun., buffalo bur. A bush-like annual to 5 dm high, densely spiny, the fruits with long spines; leaves deeply lobed and wrinkled, spiny; flowers yellow. Disturbed soil of unused areas or fields. East of a line from Montana to Arizona and Mexico.

S. tuberosum L., common potato. Annual from the tuber (potato), up to 5 dm high, leaves compound, flowers white. Potato poisoning in either humans or livestock usually comes from eating sunburned or sun-greened potatoes or the sprouts. Spoiled potatoes should never be fed to livestock.

Other species of less importance found in the central states include: *S. dimidiatum* Raf. *(S. torreyi* Gray), western horse nettle; *S. dulcamara* L., bittersweet, a woody vine usually escaped; *S. triflorum* Nutt., cut-leaved nightshade, a low-growing plant found throughout the central states and westward; *S. villosum* Mill., hairy nightshade, resembling *S. americanum* but hairy; *Nicotiana tobacum* L., cultivated tobacco, contains the alkaloid nicotine.

POISONING: A number of alkaloids, including solanine, are responsible for the poisoning. No matter which alkaloid is present in a species, the symptoms and effects are quite similar to the others. All parts of the plants are poisonous under certain conditions, even though the thoroughly ripened fruit of some species is edible.

SYMPTOMS: In humans, headache, lowered temperature, stomach pains, vomiting, diarrhea, dilated pupils, loss of sensation, depression of circulation and respiration, and death resulting from paralysis. In livestock, salivation, nausea, trembling, difficult breathing, constipation or diarrhea, and nervous stimulation followed by depression and paralysis. Death often comes too quickly for medication.

Solanum americanum flower

96

Solanum americanum leaf

Solanum carolinense leaf

Solanum americanum fruit

Solanum carolinense flower

Solanum carolinense flower

Solanum carolinense seed

Solanum carolinense fruit

97

Solanum elaeagnifolium flower

Solanum elaeagnifolium plant

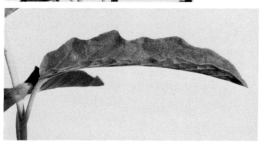

Solanum elaeagnifolium leaf

Solanum elaeagnifolium seed

Solanum rostratum bur

98

Solanum elaeagnifolium fruit

Solanum tuberosum
flower

Solanum tuberosum
sprouts

Solanum rostratum
flower

Lantana camara L.   VERBENACEAE
Lantana

A low perennial shrub, occasionally becoming vine-like in greenhouses or tropical climates. Leaves opposite, ovate, 5–10 cm long, 3–6 cm wide, toothed, aromatic, rough, hairy; leaf stalks 1–2 cm long. Flowers in axillary umbels 4–6 cm across, the stalk elongating as flowering proceeds; each flower tubular, 1–1.3 cm long, the top flare 6–8 mm across, white, pink, or yellow. Fruit berry-like, 6–8 mm diameter, black, glossy. Seeds oval, 4–5 mm long, 3–3.5 mm thick, grooved on two sides, brown, rough, irregular surface.

Grown as a potted plant in the central states, occasionally as an outdoor planting. Escaped in subtropical areas of the United States, where it becomes accessible to farm animals.

POISONING: Lantadene A is found in the leaves and fruits. Children have been poisoned fatally from eating the berries, and livestock have been poisoned from eating the leaves and berries.

SYMPTOMS: In livestock and humans, stomach and intestinal disorders accompanied by bloody feces, muscular weakness, severe circulatory trouble, and death.

99

Lantana camara flower

Lantana camara flower

Lantana camara fruit   Lantana camara seed

Lantana camara fruit, leaf

Salvia reflexa Hornem.   LAMIACEAE
Annual sage, lance-leaved sage

Herbaceous annual 2–5 dm high, few or many branches, stem angled, finely hairy. Leaves opposite, lanceolate to lance-linear, 3–5 cm long, often folded lengthwise and recurved at the end, margin entire or with a few small teeth, gray-green. Flowers in terminal racemes 5–10 cm long, 8–20 flowers, usually 2 at each node; flower irregular, 5–10 mm long, upper lip short and arched, lower lip spreading, pale blue; calyx ribbed, about 5 mm long, 2 lips; June–September. Calyx enlarges slightly at maturity and holds the seeds. Seeds oval, 2–2.5 mm long, usually 2 flattened surfaces and a rounded side, light brown with a few, fine, dark brown mottlings.

Dry prairies, roadsides, unused areas, either sandy or rocky limestone soils. Wisconsin to Montana, south to Utah, Mexico, Texas, and Arkansas. Occasionally eastward.

POISONING: The poisonous principle is unknown, but in Australia where sheep have been poisoned, the plant has a high nitrate content and the symptoms were similar to those of nitrate poisoning. In Wyoming, cattle were poisoned from eating the plant but did not show the regular symptoms of nitrate poisoning.

SYMPTOMS: Weakness and inflammation of the gastrointestinal tract.

Salvia reflexa plant

Salvia reflexa leaves

Salvia reflexa flowers

Salvia reflexa flowers

Salvia reflexa seed

Salvia reflexa fruit

101

Digitalis purpurea L.
Foxglove, digitalis

SCROPHULARIACEAE

Herbaceous perennial or biennial, 1–1.5 m tall, stem simple or branched. Leaves alternate, wrinkled, densely fine hairy, toothed; basal leaves ovate, 10–12 cm long; leaf stalks 8–10 cm long, winged; stem leaves oblanceolate, 5–8 cm long. Flowers in racemes 20–30 cm long; flowers tubular, 4–5 cm long, pendant, purple or white with dark spots at the bottom of the tube. Outer end 5-lobed, slightly flared; July–August. Fruit globose, many seeds. Seeds irregularly oval or cylindric, about 0.75 mm long, 0.5 mm wide, brown or yellow-brown, surface reticulate.

Used as a garden plant in the central states. Native of Europe, escaped along the Pacific Coast from British Columbia to California; common along roadsides, fields, and unused areas in Washington and Oregon.

POISONING: All parts of the plant contain a number of steroid glycosides which strengthen the heartbeat, but slow it down through an effect on the medulla of the brain. Poisoning of people is most common from an improper use of the drugs made from the plant. Occasionally children are poisoned from sucking the nectar from the flowers or eating the seeds. Livestock seldom eat the plant but several cases of poisoning have been reported. Drying does not destroy the poison.

SYMPTOMS: In humans, stomach disorders, diarrhea, headaches, irregular heartbeat, drowsiness, tremors, and convulsions. In addition to the above, livestock often show bloody feces and discolored urine.

Digitalis purpurea young plant

Digitalis purpurea flower

Digitalis purpurea flower

Digitalis purpurea seed

Lobelia cardinalis L.
Cardinal flower

CAMPANULACEAE

Herbaceous perennial, 1–1.5 m high, stem simple or branched at the inflorescence, erect. Leaves alternate, lanceolate to elliptic, 8–14 cm long, 2–4 cm wide, finely toothed, sessile or with short petiole. Flowering raceme terminal, 1–4 dm long, the flowers subtended by a leaf-like bract; flowers bilaterally symmetrical, 3–4 cm long, the 2 upper lobes erect, the 3 lower lobes spreading, scarlet; flower stalks hairy, 5–12 mm long; August–October. Capsules 8–10 mm long, many seeds. Seeds oval, 0.5–0.75 mm long, 0.25 mm wide, amber, rough surface.

Moist or wet ground around lakes, marshes, or streams, open or shaded areas. Quebec to Minnesota southward to the southeast corner of Colorado, and east to Texas and Florida.

Other species: *L. siphilitica* L., blue cardinal flower. A shorter, stouter plant with blue flowers, otherwise similar to *L. cardinalis.* Ranges farther west. *L. inflata* L., Indian tobacco. A hairy annual, stem usually branched, flowers blue or white, the base much inflated in fruit. Only along eastern Nebraska and Kansas in the central states but ranges from Prince Edward Island to Minnesota, south to Kansas, Mississippi, and Georgia. This is the most poisonous species. *L. spicata* Lam., pale-spike lobelia. A delicate perennial in prairies, 3–7 dm high, pale blue flowers and narrow leaves. Quebec and Minnesota south to Kansas, Texas, and Arkansas. Should be regarded with suspicion.

POISONING: The entire plant contains alkaloids such as lobelamine, lobeline, and probably many others. Early settlers learned from the Indians that these plants were used for smoking and for medicines and began to make a number of medicines to cover a variety of disorders. Human deaths resulted from misuse of the drugs.

Since the practice has virtually stopped, no deaths have been reported recently. A southwestern species, *L. berlandieri* A. DC., has been reported as poisoning cattle and goats.

SYMPTOMS: In livestock, salivation, runny nose, ulcerations, hemorrhages, and stomach disorders. In humans, nausea, weakness, tremors, and convulsions followed by death.

Lobelia cardinalis leaves

Lobelia cardinalis flowers

Lobelia cardinalis seeds

Lobelia cardinalis flowers

Lobelia cardinalis fruits

Lobelia siphilitica flowers

Lobelia siphilitica leaf

Lobelia spicata flowers

Lobelia siphilitica flowers

105

Sambucus canadensis L.
Elderberry

CAPRIFOLIACEAE

Coarse, woody shrub 2-3 m high, branches near the top, forms colonies by underground runners. Stems light brown, wood white, large pith. Leaves opposite, pinnately compound, 15-20 cm long, 5-9 leaflets; leaflets narrowly ovate, 6-12 cm long, toothed. Flowers in terminal clusters, either round topped or flat topped, clusters 8-18 cm across; flowers regular, 5-6 mm across, 5 white petals; flower stalks short and white; May-June. Fruit clusters usually drooping; berry globose, 4-6 mm diameter, deep purple, glossy, juicy, 3-4 seeds. Seeds ovoid, 2.5-3.5 mm long, 1.25 mm wide, sides rounded or with one flat side, yellow, rough with minute pits and wrinkles.

Moist soils of ditches, stream banks, fence rows, and slough margins. Nova Scotia to Manitoba, south to Kansas, Texas, and Florida.

Other species: *S. racemosa* L. ssp. *pubens* (Michx.) Koehne, red elder. A shorter plant, 1.5 m high, not forming colonies; flower and fruit clusters pyramidal, fruits bright red. Rich or rocky soil, often in rock crevices. Newfoundland to Alberta, south to Colorado, Ohio, and Georgia. Common in the Rocky Mountains and Black Hills.

POISONING: The foliage, stems, roots, and berries contain a cyanogenetic glycoside and an alkaloid; occasionally prussic acid is produced. The berries contain only small amounts of the toxins but should not be eaten raw. Cooking destroys the poison. Children have been poisoned by whistles and blow guns made from the plant's hollow stems. Cattle often eat the young foliage and pigs may dig out the roots.

SYMPTOMS: In livestock and humans, nausea, vomiting, diarrhea, and gastrointestinal pains.

Sambucus canadensis plant

Sambucus canadensis stem

Sambucus canadensis leaf
Sambucus canadensis pith

Sambucus canadensis flowers

Sambucus canadensis flowers

Sambucus canadensis
seed
Sambucus canadensis
fruit

107

Sambucus racemosus
var. pubens fruit

Sambucus racemosus var. pubens leaf

Sambucus racemosus var. pubens fruit

Ambrosia artemisiifolia L.   ASTERACEAE
Ragweed

Herbaceous annual, branched, 3-6 dm high, glabrous or hairy; slender taproot, often crooked at the crown. Leaves opposite below and alternate above, 4-9 cm long, 2-5 cm wide, 2-3 times pinnatifid, deeply cut into narrow segments. Staminate flowers green, terminal on branches, raceme slender, 2-7 cm long, erect, compact at first, becoming loose at pollination; pistillate flowers small, green, axillary just below the staminate; July-September. Fruit obconic, 3-5 mm long, with a central beak and a ring of tubercles around the top, pale brown.

Cultivated land or prairies where the sod is not solid. Sunny areas throughout the United States.

Other species: *A. psilostachya* DC., western ragweed. Resembles the above but is perennial from a deep, horizontal rhizome, usually in sandy soils of overgrazed pastures or in unused areas. Central United States and southwest into Mexico. *A. trifida* L., horse weed, giant ragweed. Tall annual, 2-3 m high; short, stout taproot; stems rough; leaves broadly ovate in outline, deeply cut into 3-5 broad lobes. Stream banks, fields, old feed lots, and farmsteads, often in shaded areas. Throughout the United States and southern Canada, not common in the northwest.

Except for a small amount of nitrates, *Ambrosia* is not a true poisonous plant, but the pollen is a main cause of hay fever in humans.

Ambrosia artemisiifolia seed

Ambrosia artemisiifolia flowers

Ambrosia artemisiifolia leaf

Ambrosia psilostachya leaves, flowers

Ambrosia artemisiifolia root

Ambrosia psilostachya rhizome

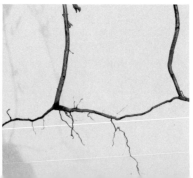

Ambrosia trifida flowers

Ambrosia trifida leaf

Ambrosia trifida seed

Eupatorium rugosum Houtt.   ASTERACEAE
White snakeroot

Herbaceous perennial 5-12 dm high, branched, stem with a few short hairs. Roots coarsely fibrous below a tough crown. Leaves opposite, simple, ovate, 5-15 cm long, 4-10 cm wide, lower edge nearly straight across, margin sharply toothed; leaf stalk about half as long as the blade. Flower clusters 4-8 cm across, composed of small "heads" 8-10 mm across, each head containing 20-30 flowers; flowers tubular, 3-4 mm across, 5 petals, white, pointed; styles elongating and extending from the flower; August-September. Fruit heads globose, 10-15 mm across; achenes narrowly obconic, 2.5-3 mm long, about 0.5 mm wide at the top, black with a pale lower end, 5-ribbed; pappus a single row of radiating, white bristles 2.5-3 mm long.

Shaded areas of moist, rich woods, stream banks, ravines, hedge rows, and thickets. Eastern Canada to Saskatchewan, south to Texas and Georgia.

Other species: No other species of Eupatorium in the central states has been involved in animal poisoning. However, Eupatorium wrightii Gray of southwest Texas, Arizona, and Mexico is a deadly poisonous plant, and an animal may die within a few hours after eating it.

POISONING: The poisonous principle is the alcohol trematol, along with glycosides. Trematol is cumulative in the animal's body.

Sickness is usually caused by eating the green plant, but it may also result from eating the hay during winter. The disease is known as "trembles" in livestock and "milk sickness" in humans. Milk from an infected cow should not be used because the poison carries over into the milk. Even suckling calves may get the disease from the milk. In the United States, "milk sickness" has been reported only in the Southeast.

SYMPTOMS: In humans, weakness, loss of appetite, abdominal pain, severe vomiting, tremors, an acrid odor on the breath, delirium, and eventual death. In livestock, trembling which increases with exercise, weakness, depression, vomiting, acrid odor on the breath, fast and difficult breathing, constipation, coma, and death.

Eupatorium rugosum leaf

Eupatorium rugosum plant

Eupatorium rugosum flowers, fruit

Eupatorium rugosum flowers

Eupatorium rugosum fruits

Eupatorium rugosum achenes

Helenium amarum (Raf.) Rock   ASTERACEAE
(H. tenuifolium Nutt.)
Sneezeweed

Herbaceous annual 2–5 dm high, branched above, glabrous, stem slightly ridged and densely covered with leaves. Leaves alternate, linear to filiform, 2–8 cm long, under 2 mm wide, often drooping; lower leaves turn brown and fall by flowering time. Flower heads on slender, naked stalks 5–7 cm long, above the leafy stem; heads 1.5–2.5 cm broad, rays 5–10 mm long, yellow; disk globose, 7–9 mm across, yellow; July–September. Achenes conical, about 1 mm long, brown with long white hairs at the base; pappus of papery scales ending in an awn.

Weedy pastures, broken areas in prairies, fence rows, and waste land, especially in sandy soils. Florida to Texas and Mexico, north to Kansas, Missouri, and Virginia.

Other species: *H. autumnale* L., bitterweed. Herbaceous perennial 5–12 dm high, branched, stem winged. Leaves elliptic to lanceolate, 5–10 cm long, 1–3 cm wide. Flowers on leafy stalks, rays 1.5–2.5 cm long, yellow; disk subglobose, yellow; August–October. Achenes silvery brown, conical, about 1 mm long, ribbed, hairy; pappus of papery scales ending in a tapered point. Moist ground of sloughs and marshes from eastern Canada to British Columbia, south to Arizona, and east to Florida. There are several other poisonous species found in the Southwest but all are similar to those found in the central area.

POISONING: The glycoside dugaldin is generally considered to
112 be the cause of poisoning, but there may be other toxins since some

animals show symptoms similar to those of aconite poisoning. Dugaldin is cumulative in the plant and is found in all parts. Milk from cows which have eaten the plant may be so bitter that it cannot be used.

SYMPTOMS: In livestock, salivation, higher pulse rate, high body temperature, difficult breathing, vomiting, weakness, convulsions, and death.

Helenium amarum
flowers

Helenium amarum plant

Helenium amarum leaves
Helenium amarum fruits

Helenium amarum flowers
Helenium amarum achenes

113

Helenium autumnale flowers

Helenium autumnale
plant

Helenium autumnale
leaves, stem

Helenium autumnale achenes

114

Helenium autumnale fruits

Hymenoxys richardsonii (Hook.) Cockll.   ASTERACEAE
Rubberweed, pingue

Herbaceous perennial, 15–30 cm high, many stems from the crown, each branched at the inflorescence. Root thick, woody, topped by a crown of old stem bases. Leaves alternate, deeply divided into narrow, linear segments about 2 mm wide; basal leaves 6–10 cm long, stem leaves 3–6 cm long, often linear and not divided; finely hairy, the base woolly. Flower heads terminal on the upper branches; heads 2–3 cm across; rays yellow, drooping or spreading, lobed at the tip, disk yellow; July–September. Achenes conical, silvery, appressed hairy; pappus scales papery, about 1.5 mm long, ovate with a long tip.

Overgrazed pastures, unused areas, roadsides on the plains or in the mountains, usually in rocky or gravelly soil. North Dakota and western Canada, south to Oregon, California, Texas and Colorado.

Other species: *H. odorata* DC., bitterweed. Similar to *H. richardsonii* but is an annual; usually only one stem which is branched near the base. Flowers in May and June, the plant dying by August. There are several other species of *Hymenoxys* in the central states but they have not been reported as poisonous.

POISONING: The toxic principle is unknown but is present in all parts of the plant, fresh or dry. The plants have a strong, bitter taste and are eaten by livestock only when other forage is inadequate. The poison is cumulative and is not destroyed by drying. Drought conditions may cause the plant to accumulate more poison. Sheep are more susceptible to being poisoned than are cattle or horses, and ingesting 1–2 percent of the animal's weight may be lethal.

SYMPTOMS: Salivation, often with a green stain around the mouth, abdominal pain, weakness and slowness of movement, depression, loss of appetite, and prostration.

Hymenoxys richardsonii
plant

Hymenoxys richardsonii
flowers

115

Hymenoxys richardsonii leaves

Hymenoxys richardsonii achenes

Hymenoxys odorata plant

Hymenoxys richardsonii fruits

Hymenoxys odorata flowers

Hymenoxys odorata leaves

116

Hymenoxys odorata fruits

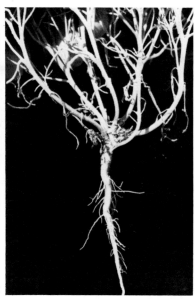

Hymenoxys odorata root

Rudbeckia laciniata L.   ASTERACEAE
Goldenglow, coneflower

Herbaceous perennial 1-3 m high from a slightly woody rootstock; stem branched, glabrous. Leaves alternate, ovate in outline, 15-20 cm long; the lower leaves with 3 divisions, each deeply cleft and toothed; upper leaves smaller, divided, or merely lobed. Flower heads on long stalks, each head subtended by green bracts; complete flower head 7-13 cm across; rays yellow, 3-6 cm long, spreading or drooping; central disk rounded, yellow; July–September. Fruiting heads hemispherical, yellow-brown. Achenes lanceolate, 4-6 mm long, 1-1.5 mm wide, 4-5 angles, dark brown.

Rich, moist soils, open or shaded. Quebec to Montana, south to Arizona, Texas, and Florida. Cultivated, often in the "double" form.

Other species: *R. hirta* L., black-eyed Susan. A native species often listed as poisonous but without substantiation. Annual, 3-6 dm high, densely hairy; flowers 4-6 cm across, disk dark brown. Prairies and unused areas throughout the United States, southern Canada, and northern Mexico.

POISONING: The poisonous principle is unknown but apparently is contained in all above-ground parts. Most of the reports of poisoning from this plant have come from the central states.

SYMPTOMS: Lack of coordination, abdominal pain, dullness, aimless wandering, and increased rate of respiration. Death occurs only in extreme cases.

Rudbeckia laciniata leaf

Rudbeckia laciniata flower

Rudbeckia laciniata fruits

Rudbeckia laciniata achenes

Rudbeckia laciniata flower

Rudbeckia hirta plant

Rudbeckia hirta flower

Senecio spp.   ASTERACEAE
Ragworts

There are many species of *Senecio* in the central states but not all are poisonous. In addition, those that are toxic do not all contain the same poisoning agent. And the whole subject of senecio poisoning is further complicated by the fact that plants of a poisonous species are not always of the same toxicity. Finally, different animals are affected in different ways by the same toxin. The species described below are the ones usually considered to be most troublesome. All except *S. jacobaea* grow in the central states.

S. douglasii DC. var. *longilobus* (Benth.) L. Benson (*S. longilobus* Benth.), woolly groundsel. Subshrubby perennial 3-5 dm tall from a woody crown, the whole plant densely white pubescent; several stems from one base, each leafy the whole length and branched at the inflorescence. Leaves alternate, 5-9 cm long, often divided into 3-5 linear segments. 3-10 flower heads at the stem summit; rays 10-12 mm long, yellow. Fruiting heads globose; achenes cylindric, 3.5-4 mm long, lightly ribbed, dark brown, appressed hairy; pappus a dense ring of silky white hairs 9-11 mm long. Prairies in sandy soil. Utah south to Mexico, west to Texas, Colorado, and southwest Kansas. Toxin is longilobine.

S. integerrimus Nutt. Suffruticose perennial 5-7 dm tall; stems single, hollow. Leaves lanceolate, hairy; basal leaves 7-10 cm long, stem leaves smaller, margin entire, undulate. Inflorescence terminal, nearly flat topped, fairly compact; floral rays 6-9 mm long, yellow.   119

Open or shaded hillsides, moist meadows, and margins of sloughs. Minnesota to British Columbia and California, east to Colorado and Iowa. Toxin is integerrimine.

S. jacobaea L., ragwort goundsel, stinking willie. Herbaceous biennial or perennial, 3–10 dm high, pinnately divided leaves up to 2 dm long; yellow flowers. Pastures and moist meadows in southeast Canada and the northeastern United States, also on the Pacific Coast. Poisoning agents are jacobine, jacodine, and senecionine.

S. riddellii T. & G., Riddell's groundsel. Herbaceous perennial with several stems from one base, 3–5 dm tall, glabrous, bright yellow-green. Flower heads numerous, rays yellow, 10–15 mm long. Prairies and roadsides. South Dakota and Wyoming, south to Arizona and Texas. Poisonous principle is riddelliine.

S. spartioides T. & G., broom groundsel. Bushy perennial, 4–6 dm high; stems somewhat woody at the base, leafy throughout. Leaves linear, entire or lobed, 5–8 cm long. Inflorescence open, many heads; rays yellow, 1–1.5 cm long. Prairies, pastures, and roadsides. South Dakota and Wyoming south to Arizona and Texas. Poisonous principle is seneciphylline.

The *Senecios* have long been associated with sickness of farm animals. A number of diseases have been named because the plants are of world-wide distribution and the toxins vary greatly. Among these diseases are "stomach staggers" in horses; Pictou in cattle; Winton disease in horses and cattle; Molteno disease in cattle; dunziekte in horses; sirasyke in horses; walking disease in horses; zdar disease in horses; and bread poisoning and venous occlusive disease in humans.

POISONING: Regardless of the name given to the separate toxins, they all belong to the pyrrolizidine alkaloids. They are not destroyed by drying, and investigators disagree as to whether they are destroyed by the fermentation in silage. Acute poisoning is not common on the range and chronic poisoning usually acts slowly— the symptoms may not be seen until it is too late. The animal may not become sick for several weeks after ingesting the first plants.

SYMPTOMS: Symptoms may vary to some extent with each toxin, but most of them are the same: standing with feet spread, depression, loss of appetite, yellowish color of the mucous membranes, uneasiness, aimless walking, abdominal pain, an unpleasant odor from the skin, diarrhea, and dark urine. At this point the animal may die quietly. The other course, especially in horses, is that the horse becomes excited, runs into buildings or other objects, walks right through fences or over a bank. In some cases the horse becomes dangerously aggressive and will attack people or such objects as a car.

Humans in other countries have been poisoned by bread made from flour contaminated with *Senecio* seeds or by drinking tea made from the leaves. The human symptoms are formation of ascites, enlarged liver, abdominal pain, nausea, vomiting, headache, and emaciation.

Senecio douglasii var. longilobus
leaves
Senecio douglasii var. longilobus plant

Senecio douglasii var. longilobus flowers
Senecio douglasii var. longilobus fruits

Senecio integerrimus leaves

Senecio integerrimus plant

Senecio integerrimus fruits
Senecio integerrimus flowers

Senecio plattensis stem
leaf
Senecio plattensis flowers
Senecio plattensis basal leaf

122

Senecio riddellii plant

Senecio riddellii leaves

Senecio riddellii flowers

Senecio riddellii fruits   *Senecio riddellii* achenes

Solidago spp.   ASTERACEAE
Goldenrod

The goldenrods are commonly thought of as causing hay fever in people and are much less known for their toxic effect on livestock who ingest the leaves at flowering time. This is from a resinous irritant but reported cases have been in question for one reason or another.

123

In general, the goldenrods are similar in that they are erect, perennial plants with narrow leaves (a few have broad leaves) and a cluster of yellow flowers at the top of the stem. This cluster may be flat-topped, pyramidal, or columnar. All of them flower in late summer, and anyone subject to hay fever should take special precaution at that time. As a group, they are well distributed over the United States.

Solidago rigida flowers

Solidago canadensis flowers

Solidago petiolaris flowers

Tanacetum vulgare L.   ASTERACEAE
Tansy

Herbaceous perennial from a short, stout rhizome, stems usually in clusters of 10-50, erect, 5-10 dm high, leafy the whole length. Leaves pinnately divided, 10-20 cm long, central axis winged; leaflets lobed or sharply toothed, fern-like. Flowers in terminal, flat-topped clusters 5-15 cm across, 50-100 heads in each; heads discoid, 6-12 mm across, yellow; without rays or with rays 1-3 mm long, yellow. June-October. Achenes about 2 mm long, yellow-brown, ribbed, flower parts often remain attached.

Roadsides and creek banks, moist or dry. Native of Europe, now in scattered locations through southern Canada and the northern United States. Common in parts of the Black Hills.

POISONING: All above-ground parts contain tanacetin, an oil which is poisonous to both humans and livestock. Livestock seldom eat the plant. Human illness and fatalities are usually the result of misuse of drugs made as home remedies to induce abortion, kill intestinal worms, or sedate the nerves.

SYMPTOMS: In humans and livestock, feeble but rapid pulse, severe gastritis, and convulsions.

Tanacetum vulgare plant
Tanacetum vulgare leaves

Tanacetum vulgare leaf   125

Tanacetum vulgare flowers
Tanacetum vulgare flowers

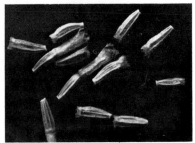

Tanacetum vulgare achenes
Tanacetum vulgare fruits

Xanthium strumarium L.　　　　　　ASTERACEAE
Cocklebur

Herbaceous annual, 3-8 dm high; stem branched, appressed hairy, rough, green marked with short black streaks. Leaves alternate, broadly ovate to suborbicular, 5-10 cm long and about as wide, base deltoid or cordate, margin toothed, with or without small lobes; leaf stalk 5-10 cm long. Flowers in small axillary clusters, green, inconspicuous; July–August. Bur densely spiny, ovoid, 15-25 mm long, 12-14 mm broad including the spines; spines hooked on the end, brown. Achenes lanceolate, 15-17 mm long, 4-5 mm wide, black, slightly ridged; seed inside the achene, brown, 12-13 mm long.

Fields, unused areas, roadsides, ditches, and pond shores. Common throughout the United States and southern Canada. Because of differences of taxonomic opinion, the name *X. strumarium* is used here to cover all the cockleburs of the central states excepting *X. spinosum* L.

POISONING: The poisonous principle is hydroquinone and is present in the seeds and seedlings. Since the plant is common in pastures, farm lots, and along the shores of farm ponds, it is readily accessible to farm animals. The burs may also be mechanically harmful to animals.

SYMPTOMS: Weakness, rapid pulse and breathing, depression, low temperature, vomiting, anorexia, lack of coordination, twitching, paralysis, inflammation of mucous membranes, and spasms. Death often occurs before the symptoms can be diagnosed.

Xanthium strumarium plants

Xanthium strumarium leaf
Xanthium strumarium flowers

Xanthium strumarium stem
Xanthium strumarium burs

Xanthium strumarium seeds

127

Triglochin maritimum L.
Arrowgrass
JUNCAGINACEAE

Herbaceous perennial from a persisting rootstock, all leaves from the base. Leaves linear, 2-4 dm long, 1-3 mm wide, grass-like but somewhat fleshy, erect. Flower stems directly from the rootstock, 3-6 dm high, naked, the raceme on the upper one-half or one-third, open or compact. Flowers small, 1-2 mm long, appressed to the stem; flower stalks 1-2 mm long, erect; July–August. Fruit cylindric, 2-3 mm long, ribbed, brown. Seeds oblong, slightly curved, 2-2.5 mm long, 0.5 mm thick, brown, smooth.

Marshy areas, creek flood plains, and wet meadows; found in wet alkaline soils, either brackish or fresh water. Alaska to Labrador, south to Delaware, Ohio, Iowa, Nebraska, Colorado, Arizona, and Mexico.

Other species: *T. palustris* L. and *T. concinnum* Davy are similar, and they are also both poisonous; a botanist should be consulted to distinguish the three species.

POISONING: All parts of the plant are poisonous due to the hydrocyanic acid content. The green plant is more poisonous than when dried with hay, but it is still dangerous in the dry state.

SYMPTOMS: Those of cyanide poisoning. Difficult breathing, twitching, trembling, spasms, and death.

Triglochin maritimum plant

128

Triglochin maritimum
rootstock

Triglochin maritimum
flowers

Triglochin maritimum
fruits

Triglochin maritimum
fruits

Triglochin maritimum
seeds

Festuca arundinacea Schreb.   **POACEAE**
(F. *elatior* L. var. *arundinacea* [Schreb.] Wimm.)
Tall fescue

Perennial grass 6–10 dm high, growing in clumps of 10–30 flower stalks, producing flowers the second year of growth. Leaves linear, 1–6 dm long, 4–9 mm wide, rough above; sheath around the stem glabrous. Flower stalks glabrous; panicle loose, 15–30 cm long, lowest branches one-third to one-half as long as the whole panicle; flower "chaff" often with short awns; June–July.

129

Unused areas, roadsides, farmsteads, cultivated for pasture and often escapes. Throughout most of the United States.

POISONING: Poisoning agent is unknown and there is some question as to whether the poison is in the plant or in an unidentified fungus growing on it. All parts of the plant, whether green or dry, may contain the poison at any time of the year. The disease, which is restricted mainly to cattle, is called "fescue foot" and may appear in one pasture and not in another, or it may appear and then disappear in the same pasture.

SYMPTOMS: Shivering, lameness, swelling in the legs, necrosis, gangrene, loss of weight, and a possible sloughing of the tail.

Festuca arundinacea plant

Festuca arundinacea fruits

Festuca arundinacea flower head

Festuca arundinacea flower head

Hordeum jubatum L.   POACEAE
Squirreltail, foxtail barley

Perennial grass 3-7 dm high, stems clustered, erect or decumbent at the base. Leaves 10-25 cm long, 2-5 mm wide, rough to the touch. Flowers in a close spike 4-8 cm long not including the awns, erect at first and then nodding; awns 3-6 cm long; June–August.

Squirreltail is not a poisonous plant but is listed here because of its physical damage to animals. The long awns are minutely barbed and once they puncture a tissue, the animal's movements cause the awn to work inward. It is particularly dangerous in hay where the animal forces its nose into the feed. Awns enter the flesh of the mouth, nose, and around the eyes, causing swelling and infection. This creates trouble in eating, breathing, or seeing, and death may result in extreme cases.

Other species: *H. pusillum* Nutt., little barley. A close relative with shorter awns, and the stems often single from the base. *H. vulgare* L., common barley. Barley may contain toxic quantities of nitrates as well as toxins from a fungus on the grain. Other genera of grasses with long awns are: *Hystrix, Setaria, Stipa, Aristida,* and *Elymus.*

Hordeum jubatum plant

Hordeum jubatum flower head

Hordeum jubatum fruits   131

Sorghum halepense (L.) Pers.   POACEAE
Johnson grass

Herbaceous, coarse perennial grass from creeping rhizomes and tangled, matted rootstocks; difficult to eradicate because each broken section of the rhizome may start a new plant. Stems generally glabrous, 1–2.5 m high, leafy the whole length. Lower leaves 4–8 dm long, 1–2 cm wide, the lower portion forming a sheath around the stem. Flower panicles terminal, loose, pyramidal, 2–4 dm long; individual flowers small. July–September. Seeds enclosed in loose bracts (glumes), ellipsoid, about 4 mm long, glossy, slick, yellow to brown.

Open areas, along creek banks, farm lots, roadsides, and fields. Across the southern half of the United States as far north as Iowa and Nebraska.

Related forms: milo, kafir, feterita, sudan grass, broom corn, and the weedy "wild cane." These are all cultivated annuals and may be involved in poisoning. They may escape but do not persist.

POISONING: The poisoning agent is dhurrin, a cyanogenetic glycoside. Some varieties of sorghum accumulate more cyanide than others, and, even within one species, other conditions influence the amount of poison. Drought and an early frost may cause high cyanide content. High nitrogen and low phosphorus content of the soil are also factors in the increased accumulation of cyanide. Young or second growth plants are especially dangerous, with the cyanide content being greatest just before pollination. Nitrate poisoning and photosensitization have also been reported from animals eating the sorghums.

SYMPTOMS: Salivation, increased and difficult breathing, dizziness, rapid and weak pulse, lack of coordination, spasms, and coma.

Sorghum halepense plant

Sorghum halepense flower head

132

Sorghum halepense fruits
Sorghum halepense rhizomes

Sorghum vulgare seeds

Sorghum vulgare fruit head
Sorghum bicolor seeds

Arisaema triphyllum (L.) Schott   ARACEAE
Jack-in-the-pulpit

    Herbaceous perennial from a globose corm about 3 cm diameter; flower and leaf stems directly from the underground corm; the base of the leaf stalk surrounds the base of the flower stalk; leaf stalks 30–50 cm long. Usually 2 leaves, each with 3 leaflets 15–25 cm long;   133 terminal leaflet elliptic to ovate, the laterals uneven at the base, occasionally whitish beneath. Flower stalk 15–30 cm high, terminated by a narrow, obconic spathe, the "hood" tapered to a point and arched over the flowers inside the spathe body; the outside of the spathe usually green, the inside purple. Flowers on an erect spike within the spathe, the staminate above the pistillate, each flower small; April–June. Fruits bright red, in a cylindric cluster 4–6 cm long, 1–2 seeds in each fruit. Seeds ovoid or globose with one flattened surface, 4–5 mm long, light red-brown, surface minutely wrinkled.

Rich, moist soils of wooded areas or on recently cleared land. East of a line from eastern North Dakota to eastern Texas.

POISONING: The whole plant contains minute crystals of calcium oxalate, the corm being the most dangerous. These crystals quickly enter the mouth tissues and cause swelling and possibly choking. A person seldom takes more than the first taste and there are no reports of death. Livestock are not affected.

SYMPTOMS: Extreme burning of the mouth and throat.

Arisaema triphyllum plant

Arisaema triphyllum flower inside

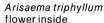

Arisaema triphyllum flower

134

Arisaema triphyllum
corm

Arisaema triphyllum
fruits

Arisaema triphyllum
seeds

Dieffenbachia seguine Schott   ARACEAE
Dumbcane, dieffenbachia

A potted plant in the central states, grown outdoors in the southern states and Hawaii. Stem stout, green, transversely marked with light-colored leaf scars. Leaves oblong, 3-6 dm long, 1-2 dm wide; solid green or variously mottled or streaked with yellow-green; leaf stalks winged. Flowers in a spike enclosed in a spathe 20-25 cm long and 2-3 cm thick, erect, drooping, or arched.

POISONING: Calcium oxalate crystals are present in all parts of the plant. If eaten, these cause a stinging and burning effect in the mouth and throat. In addition, the stem contains a toxin which has not been completely identified, possibly a protein.

SYMPTOMS: Severe stinging and burning in the mouth and throat; at times the tongue becomes immobile, thus the name dumbcane. The toxin causes vomiting, diarrhea, and swelling of the mouth and throat.

135

Dieffenbachia seguine
stem

Dieffenbachia seguine
plant

Dieffenbachia seguine flower

Dieffenbachia seguine plant

Philodendron spp.   ARACEAE
Philodendrons

The common name philodendron is applied loosely to a number
of species or genera of plants resembling each other. Since they are
similar and so little is known about their actual toxicity, they are
discussed here as a group.

In the central states they are all potted plants and are among the

136

most commonly used ornamentals. They may be short, bush-like plants or vines trailing for several meters, often trained over fireplaces or windows. The leaf shape varies considerably but all are dark green and glossy. The most common philodendron has ovate, entire-margined leaves 8–15 cm long, 5–8 cm wide. Others have leaves the shape of a triangular heart, 3–5 dm long and 2–3 dm wide. *Monstera,* often called philodendron, has deeply dissected leaves 2–4 dm long and 1.5–3 dm wide. The flowers are characteristic of the *Araceae,* a sheath (spathe) surrounding the dense spike of flowers (spadix).

In addition to these, such plants as *Caladium* and *Colocasia* (elephant's ear) are commonly grown in pots or gardens. They must be regarded with suspicion as possibly poisonous, although the tuber of *Colocasia esculenta* Schott is edible.

POISONING: The leaves and stems contain needle-like crystals of calcium oxalate which cause a stinging and burning of the mouth and throat. Humans and other animals have been harmed in this way. It is quite possible that a toxin is also involved because there are cases on record of cats eating the leaves of philodendron and about half of these cases ended in death. One case is on record of cattle becoming quite sick from eating the leaves.

SYMPTOMS: In humans, farm animals, and pets, stinging and burning of the tissues of the mouth and throat. Laziness, weakness, and loss of kidney function.

Philodendron plant

Philodendron plant

Philodendron plant

Monstera flowers

Monstera plant

Allium canadense L.   LILIACEAE
var. *lavandulare* (Bates) Ownbey
Wild onion, garlic

Herbaceous perennial 3-7 dm high from an ovoid or globose bulb 10-15 mm across; outer bulb coat brown, fibrous, inner layers white. Leaves basal or near the base, 2-3.5 dm long, 1-5 mm wide, flattened with rounded or keeled back. Flower stem 2-6 dm high, flowers clustered at the top and subtended by 3 papery bracts; flowers with 6 petaloid parts, white or pink; May–June. 3-4 seeds from each flower; seeds generally ovoid, about 2 mm long, black. In variety *canadense,* the flowers are replaced by small bulblets.

Prairies, unused areas, and open woods, occasionally in moist soils. East of a line from North Dakota to Texas and north into southeastern Canada.

Other species: *A. cepa* L., cultivated onion. This plant has caused poisoning when cattle or horses were pastured in a field where the culled onions remained. Experimental testing has shown that dogs were poisoned from eating onions added to their diet. *A. schoenoprasm* L., chives. This has caused poisoning in horses in early spring when the plant was abundant and other forage absent.

POISONING: The toxin has not been thoroughly identified but has been reported as an alkaloid. It is present in both the leaves and bulbs.

SYMPTOMS: Anemia, a jaundice condition, hemoglobinuria, and the odor of onions on the breath. Death may result in a few days.

Allium canadense flowers

Allium canadense seeds

Allium canadense bulblets

Allium canadense bulb

Allium cepa plant

Allium sativum leaves

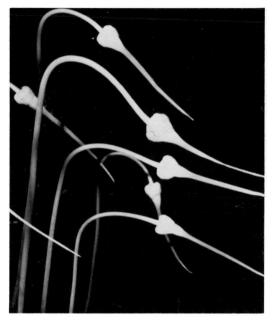

Allium sativum bulb

Allium sativum flower heads

Convallaria majalis L.   LILIACEAE
Lily of the valley

Herbaceous perennial from a thick, vertical rootstock and a deep rhizome, forming dense colonies; plant 15–25 cm tall, glabrous. Leaves broadly elliptic, 10–20 cm long, 3–7 cm wide, parallel veins, base of leaf sheathing the flowering stem, usually 2 leaves, erect or arched. Flower racemes 6–10 cm long, 5–10 flowers on drooping stalks, total height of flower stem 12–20 cm; flowers white, urn-shaped, about 7 mm long and 8 mm wide, the 6 small lobes recurved, flower fragrant; April–May. Fruit globose, 12–15 mm long, 10–12 mm thick, orange-red when mature; 6–9 seeds. Seeds irregularly ovoid, 4–5 mm long, 3–4 mm wide, light yellow-brown.

Lily of the valley was introduced from Eurasia and is now planted in moist, shaded areas throughout the United States. The plant is aggressive and will cover an area to the exclusion of other plants. It will also persist for several years after abandonment but seldom escapes to new areas. A native plant, *C. montana* Raf., is found in the southeastern United States and is considered dangerous.

POISONING: The cardiac glycosides convallarin and convallamarin are found in all parts of the plant. Although no deaths of either humans or animals have been reported, these toxins have an effect on the heart similar to that of digitalis.

SYMPTOMS: Irregular heartbeat and stomach trouble.

Convallaria majalis colony
Convallaria majalis plant

Convallaria majalis
flowers

Convallaria majalis fruits

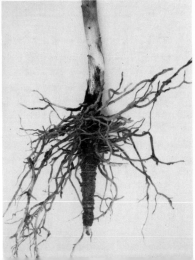

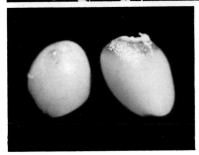

Convallaria majalis seeds

Convallaria majalis rootstock

Hyacinthus orientalis L.   LILIACEAE
Hyacinth

Herbaceous plant 20–35 cm high arising from a bulb 3–5 cm diameter, the bulb covered with brown, skin-like layers. Leaves narrow, parallel veins, 20–30 cm long, 2–3.5 cm wide, trough-shaped, the tip slightly cupped; all leaves from ground level. Flower racemes usually overtopping the leaves, raceme dense with white, blue, or pink lily-like flowers; flowers 2 cm long, 2.5–3 cm across the flared end, the petaloid parts recurved. Early spring. Fruits ovoid or globose, about 1 cm across, 3 divisions. Seeds ovoid, 3–4 mm diameter, black, minutely pitted.

The hyacinth is well known both in early spring as a potted plant and outdoors a little later.

POISONING: The poisonous principle is not known but all parts of the plant may cause sickness if eaten. The bulb contains the greatest amount of toxin. Both humans and livestock have been poisoned from eating some part of the plant.

SYMPTOMS: Stomach ache and cramps, accompanied by vomiting and diarrhea.

Hyacinthus orientalis plant

Hyacinthus orientalis fruit

142

Hyacinthus orientalis flower

Hyacinthus orientalis seed

Hyacinthus orientalis bulb

Zigadenus elegans Pursh   LILIACEAE
White camass

Herbaceous perennial from a bulb. Stem simple, 3–7 dm high, glabrous, few small leaves on it. Bulb ovoid, 2–3 cm long, 1.5–2 cm thick, similar to an onion, outer coat dark brown. Leaves linear, mostly basal, 2–3 dm long, 8–14 mm wide, trough-shaped, glabrous, margin entire. Flowers in open racemes, 1–2 dm long; flowers regular, 13–20 mm across, the 6 petaloid parts pale yellow or white with a yellow center; June–July. Capsules narrowly conic, 13–20 mm long, opening at the tip, many seeds. Seeds 5–6 mm long, rounded ends, yellow-brown, appear like a piece of crumpled paper.

Prairies, meadows, and open woods. Alaska to Manitoba and Minnesota, south to Missouri, Texas, and Arizona. Not on the West Coast.

Other species: *Z. venenosus* Wats., death camass. A smaller plant with narrower leaves; the same habitat but the range is more northern and extends to the West Coast. *Z. nuttallii* Gray, death camass. A coarser plant than *Z. elegans;* leaves 3–4 dm long, yellow-green; bulb 3–5 cm long; flowers pale yellow. Prairies from Tennessee to eastern Nebraska, Kansas, and Texas.

POISONING: All parts of the plant contain several steroid alkaloids of the veratrum group, among which is zygacine. Children have been poisoned from eating the bulb or chewing the flowers; livestock by eating the green or dry leaves and possibly the bulb.

SYMPTOMS: Salivation, nausea, vomiting, weakness, trembling, staggering, low temperature, difficult breathing, and coma. In children an upset stomach is one of the first symptoms.   143

Zigadenus elegans flower
Zigadenus elegans seed

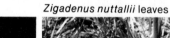

Zigadenus elegans flower
Zigadenus nuttallii leaves

Zigadenus nuttallii plant

144   *Zigadenus nuttallii* flowers

Zigadenus nuttallii fruits

Zigadenus nuttallii bulb

Zigadenus venenosus flowers

Zigadenus venenosus bulb

Zigadenus venenosus plant

Zigadenus venenosus fruits

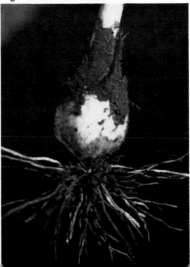

145

Iris missouriensis Nutt.
Wild iris

IRIDACEAE

Herbaceous perennial from a thick, horizontal rhizome near the surface of the ground, rhizome covered with old leaf bases and papery scales; coarse roots extending from the rhizome. Principle leaves from the rhizome, linear, 2-6 dm long, 1-2 cm wide, pointed, light green, parallel veins, smaller leaves on the flower stalk. 1-4 flowers near the summit of the stalk which is usually 3-4 dm long but up to 8 dm in marshy areas; flowers similar to those of the cultivated iris but more delicate, 5-8 cm long; calyx blue with darker veins and tinged with yellow at the base, recurved; petals light blue, erect. Capsule oblong, 3-5 cm long, 1-1.5 cm thick, somewhat 3-sided, indented between the seeds. Seeds ovoid, about 4 mm long, 2.5-3 mm wide, dark brown, dull, minutely wrinkled.

Roadsides, open woodlands, and meadows either in mountainous areas or on the plains. North Dakota to British Columbia, south to California and New Mexico, northeast to Colorado and western Nebraska. Common in the Black Hills. *Iris versicolor* L. is the common wild iris of the eastern states. The common garden iris is also dangerous.

POISONING: Most of the genus *Iris* contain irisin, an acrid resinous substance which causes illness. It is most abundant in the rootstock but there are sufficient quantities in the leaves and flowers to cause trouble in livestock when eaten in large quantities. Handling the rootstock may cause dermatitis in some people.

SYMPTOMS: In livestock and humans, gastric disturbances and difficult breathing, occasionally purgation.

Iris missouriensis plants

Iris missouriensis flower

146

Iris missouriensis fruit

Iris missouriensis fruit, leaf

Iris missouriensis seed

Iris missouriensis rootstock

Garden iris flower

Garden iris plant

147

POISONOUS PLANTS
BY TOXIC PRINCIPLE

Toxin	Scientific Name	Common Name	Toxic Part
ALKALOIDS			
	Aconitum spp.	monkshood	all parts
	Allium spp. (probable)	onion, chives	all parts
	Amaryllis belladonna L.	amaryllis	bulbs
	Argemone spp.	prickly poppy	all parts
	Asclepias spp.	milkweed	all parts
	Atropa belladonna L.	belladonna	all parts
	Baptisia spp.	false indigo	all parts
	Caulophyllum thalictroides (L.) Michx.	blue cohosh	all parts
	Colchium autumnale L.	autumn crocus	all parts
	Conium maculatum L.	poison hemlock	all parts
	Corydalis spp.	fitweed	all parts
	Crotalaria spp.	rattlebox	all parts
	Datura spp.	Jimsonweed	all parts
	Delphinium spp.	larkspur	all parts
	Dicentra spp.	Dutchman's breeches, bleeding heart	all parts
	Echium vulgare L.	viper's bugloss	all parts
	Festuca arundinacea Schreb.	tall fescue	tops
	Fritillaria spp.	fritillary lily	all parts
	Gloriosa superba L.	gloriosa lily	all parts
	Hyoscyamus niger L.	henbane	all parts
	Laburnum anagyroides Medic.	golden chain	flowers, seeds
	Lobelia spp.	lobelia	all parts
	Lolium spp.	darnel, rye grass	tops
	Lupinus spp.	lupine, bluebonnet	all parts
	Lycopersicon esculentum Mill.	tomato	leaves
	Menispermum canadense L.	moonseed	leaves, fruits
	Nicotiana spp.	tobacco	tops
	Narcissus spp.	narcissus, jonquil, daffodil	bulbs
	Ornithogalum umbellatum L.	star of Bethlehem	all parts
	Papaver spp.	poppies	leaves, seed
	Robinia pseudoacacia L.	black locust	all parts
	Sambucus spp.	elderberry	all parts
	Sanguinaria canadensis L.	bloodroot	all parts
	Senecio spp.	groundsel	all parts
	Solandra spp.	trumpet flower	all parts
	Solanum spp.	nightshade, potato	leaves, stems, tuber if sun-greened
	Thermopsis spp.	golden pea	all parts

149

Toxin	Scientific Name	Common Name	Toxic Part
	Taxus spp.	yew	leaves, seeds
	Zigadenus spp.	camass	all parts
DERMATITIS; HAY FEVER			
	Ailanthus altissima (Mill.) Swingle	tree of heaven	leaves, flowers
	Ambrosia spp.	ragweed, horseweed	leaves, pollen
	Anagallis arvensis L.	scarlet pimpernel	leaves
	Anthemis cotula L.	dog fennel	leaves, flowers
	Arisaema triphyllum (L.) Schott	Jack-in-the-pulpit	all parts
	Asarum canadense L.	wild ginger	leaves
	Asimina triloba (L.) Dun.	pawpaw	fruit, leaves
	Campsis radicans (L.) Seem.	trumpet creeper	leaves, flowers
	Cannabis sativa L.	marijuana	leaves
	Catalpa spp.	catalpa	flowers
	Caulophyllum thalictroides (L.) Michx.	blue cohosh	roots
	Chrysanthemum spp.	chrysanthemum, oxeye daisy	leaves
	Clematis virginiana L.	virgin's bower	leaves
	Conyza canadensis (L.) Cronq. (Erigeron canadensis)	mare's tail	leaves
	Datura stramonium L.	Jimsonweed	leaves, flowers
	Delphinium spp.	larkspur	leaves
	Dicentra spp.	Dutchman's breeches, bleeding heart	all parts
	Echium vulgare L.	viper's bugloss	leaves, stems
	Euphorbia spp.	spurge	milky juice
	Ginkgo biloba L.	maidenhair tree	fruit
	Helleborus niger L.	Christmas rose	leaves
	Humulus lupulus L.	hops	leaves
	Hypericum perforatum L.	St. Johnswort	leaves
	Iris spp.	iris	rhizomes
	Juniperus virginiana L.	juniper, cedar	leaves
	Laportea canadensis (L.) Wedd.	wood nettle	stinging hairs
	Leonurus cardiaca L.	motherwort	leaves
	Maclura pomifera (Raf.) Schneid.	osage orange	milky juice
	Mentzelia oligosperma Nutt.	stickleaf	hairs
	Pastinaca sativa L.	wild parsnip	all parts
	Phacelia spp.	phacelia	leaves
	Podophyllum peltatum L.	mayapple	roots
	Polygonum spp.	smartweed	leaves, sap
	Ranunculus spp.	buttercup	leaves
	Rumex spp.	dock	leaves
	Sanguinaria canadensis L.	bloodroot	sap
	Solidago spp.	goldenrod	pollen
	Toxicodendron spp.	poison ivy, poison oak	all parts
	Tragia spp.	noseburn	stinging hairs
	Urtica dioica L.	stinging nettle	stinging hairs
GLYCOSIDES, CARDIAC			
	Adonis vernalis L.	pheasant's eye	all parts
	Apocynum ssp. (suspected)	dogbane	all parts
	Convallaria majalis L.	lily of the valley	all parts
	Digitalis purpurea L.	foxglove	all parts
	Nerium oleander L.	oleander	all parts
GLYCOSIDES, COUMARIN			
	Aesculus spp.	buckeye	all parts
	Melilotus spp.	sweet clover	stem, leaves
CLYCOSIDES, CYANOGENETIC			
	Cercocarpus montanus Raf.	mountain mahogany	leaves

150

Toxin	Scientific Name	Common Name	Toxic Part
	Glyceria striata (Lam.) Hitchc.	fowl mannagrass	tops
	Holcus lanatus L.	velvet grass	all parts
	Hydrangea spp.	hydrangea	leaves
	Linum spp.	flax	leaves, seeds
	Lotus corniculatus L.	birdsfoot trefoil	all parts
	Phaseolus lunatus L.	lima bean	all parts
	Picradeniopsis spp. *(Bahia)*	bahia	all parts
	Prunus spp.	cherry, peach, apricot, almond	bark, leaves, seeds
	Pyrus malus L.	apple	seeds
	Sambucus spp.	elderberry	all parts
	Sorghum spp.	sudan, Johnson grass, milo, etc.	all parts
	Suckleya suckleyana (Torr.) Rydb.	poison suckleya	all parts
	Trifolium repens L.	white clover	tops
	Triglochin spp.	arrowgrass	all parts
	Vicia sativa L.	spring vetch	tops
GLYCOSIDES, GOITROGENIC			
	Beta vulgaris L. var. *cicla* L.	chard	all parts
	Brassica spp.	kohlrabi, kale, broccoli, cabbage, Brussels sprouts, rutabaga, Chinese cabbage, turnip (root), rape (seed), white mustard (seed), black mustard (seed)	most parts
	Glycine max (L.) Merr.	soybean	improper extraction of oil
	Linum usitatissimum L.	flax	leaves and seed
GLYCOSIDES, MUSTARD OILS			
	Amoracia lapathifolia Gilib.	horseradish	all parts
	Barbarea vulgaris R. Br.	yellow rocket	all parts
	Brassica spp.	white mustard, rape, cabbage, kale, turnip, Indian mustard, broccoli, charlock	all parts
	Erysimum chieranthoides L.	wormseed wallflower	all parts
	Raphanus sativa L.	wild radish	all parts
	Thlaspi arvense L.	pennycress, fanweed	seeds
GLYCOSIDES, PROTOANEMONIN			
	Actaea spp.	baneberry	all parts
	Anemone spp.	windflower	all parts
	Caltha palustris L.	marsh marigold	all parts
	Ranunculus spp.	buttercup	all parts
GLYCOSIDES, SAPONIN			
	Agrostemma githago L.	corn cockle	seeds
	Baccharis spp.	baccharis	new growth
	Gutierrezia spp.	broomweed	all parts
	Hedera helix L.	English ivy	leaves
	Medicago sativa L.	alfalfa	tops
	Phytolacca americana L.	pokeberry	all parts
	Saponaria officinalis L.	bouncing bet	all parts
	Sesbania spp.	coffeeweed	seeds
MECHANICAL			
	Anemone patens L.	Pasque flower	hair balls
	Aristida spp.	poverty grass	awns
	Bidens spp.	Spanish needle	fruits
	Cenchrus spp.	sandbur	burs
	Centaurea spp.	star thistle	spines
	Elymus spp.	rye grass	awns

151

Toxin	Scientific Name	Common Name	Toxic Part
	Hordeum jubatum L.	squirreltail	awns
	Hystrix patula Moench	bottlebrush	awns
	Pisum sativum L.	pea (hay)	impaction
	Rubus spp.	blackberry	thorns
	Setaria spp.	foxtail	awns
	Solanum rostratum Dun.	buffalo bur	burs
	Stipa spp.	needle grass	seeds
	Trifolium spp.	clover	hair balls
	Xanthium spp.	cocklebur	burs
NITRATES			
	Amaranthus spp.	pigweed	all parts
	Amsinckia spp.	tarweed	leaves
	Ambrosia tomentosa Nutt. *(Franseria)*	white ragweed	leaves
	Apium graveolens L.	celery	leaves
	Avena sativa L.	oats	all parts
	Beta vulgaris L.	beets	leaves
	Bidens frondosa L.	beggar ticks	all parts
	Brassica spp.	rutabaga, rape, broccoli, turnip	leaves
	Carduus spp.	nodding thistle	leaves
	Chenopodium spp.	lamb's quarters	all parts
	Cirsium arvense (L.) Scop.	Canada thistle	leaves
	Cleome serrulata Pursh	bee plant	leaves
	Conium maculatum L.	poison hemlock	all parts
	Convolvulus spp.	bindweed	all parts
	Cucumis sativus L.	cucumber	all parts
	Cucurbita maxima Duchesne	squash	all parts
	Datura spp.	Jimsonweed	all parts
	Daucus carota L.	wild carrot, Queen Anne's lace	all parts
	Echinochloa spp.	barnyard grass	all parts
	Eleusine indica (L.) Gaertn.	goose grass	all parts
	Eupatorium perfoliatum L.	Joe-pye weed	all parts
	Eupatorium purpureum L.	thoroughwort	all parts
	Glycine max (L.) Merr.	soybean	all parts
	Gnaphalium purpureum L.	purple cudweed	all parts
	Helianthus annuus L.	sunflower	all parts
	Hordeum vulgare L.	barley	all parts
	Ipomoea batatas Lam.	sweet potato	vines
	Kochia scoparia (L.) Schrad.	firebush	tops
	Lactuca sativa L.	lettuce	tops
	Lactuca serriola L. *(L. scariola)*	prickly lettuce	tops
	Linum spp.	flax	tops
	Lygodesmia juncea (Pursh) Hook.	skeleton plant	tops
	Malva parviflora L.	small mallow	tops
	Medicago sativa L.	alfalfa	tops
	Melilotus officinalis L.	sweet clover	tops
	Panicum capillare L.	panic grass	tops
	Plagiobothrys spp.	popcorn flower	tops
	Polygonum spp.	smartweed	tops
	Raphanus sativus L.	radish	all parts
	Rumex spp.	dock	tops
	Salsola iberica Sennen & Pau *(S. pestifer)*	Russian thistle	tops
	Salvia reflexa Hornem. (probable)	annual sage	tops
	Sambucus racemosa L. ssp. *pubens* (Michx.) House	red elder	all parts
	Secale cereale L.	rye	tops

Toxin	Scientific Name	Common Name	Toxic Part
	Solanum spp.	nightshades, potato	all parts (sun-greened tubers)
	Solidago spp.	goldenrods	tops
	Sonchus spp.	sowthistle	tops
	Sorghum spp.	Johnson grass, milo, sudan grass	tops
	Stellaria media (L.) Cyr.	chickweed	tops
	Tribulus terrestris L.	puncture vine	tops
	Triticum sativum Lam.	wheat	tops
	Urtica dioica L.	stinging nettle	tops
	Verbesina encelioides (Cav.) B. & H.	golden crownbeard	tops
	Zea mays L.	corn	tops
OXALATES, crystals			
	Arisaema triphyllum (L.) Schott	Jack-in-the-pulpit	all parts
	Caladium spp.	caladium	all parts
	Calla palustris L.	wild calla	all parts
	Colocasia spp.	elephant's ear	all parts
	Dieffenbachia spp.	dumbcane	all parts
	Monstera spp.	philodendron	all parts
	Philodendron spp.	philodendron	all parts
OXALATES, soluble			
	Beta vulgaris L.	beets	tops
	Chenopodium album L.	lamb's quarters	tops
	Oxalis spp.	sorrel	tops
	Portulaca oleracea L.	purslane	tops
	Rheum rhaponticum L.	rhubarb	leaf blades
	Rumex spp.	dock	tops
	Salsola iberica Sennen & Pau	Russian thistle	tops
	Sarcobatus vermiculatus (Hook.) Torr.	greasewood	leaves
PHOTOSENSITIZERS			
	Avena sativa L.	oats	stems, leaves
	Brassica napus L.	rape	stems, leaves
	Cynodon dactylon (L.) Pers.	Bermuda grass	leaves
	Euphorbia spp.	spurge	stems, leaves
	Fagopyrum sagittatum Gilib.	buckwheat	stems, leaves
	Hypericum perforatum L.	St. Johnswort	leaves
	Kochia scoparia (L.) Schrad.	firebush	leaves
	Lantana spp.	lantana	leaves
	Medicago sativa L.	alfalfa	stems, leaves
	Panicum spp.	panic grass	leaves
	Polygonum spp.	smartweed	leaves
	Sorghum vulgare Pers.	sudan grass	stems, leaves
	Tetradymia spp.	horsebush	leaves
	Tribulus terrestris L.	puncture vine	leaves
	Trifolium spp.	clover	leaves
	Vicia spp.	vetches	leaves
PHYTOTOXINS			
	Ricinis communis L.	castor bean	all parts
	Robinia pseudoacacia L.	black locust	all parts
RESINOIDS			
	Asclepias spp.	milkweeds	tops
	Azalea spp.	azalea	all parts
	Cannabis sativa L.	marijuana	tops
	Cicuta maculata L.	water hemlock	all parts
	Iris spp.	iris	all parts
	Podophyllum peltatum L.	mayapple	all parts
	Rhododendron spp.	rhododendron	all parts
	Solidago spp.	goldenrod	leaves at flowering time

Toxin	Scientific Name	Common Name	Toxic Part
SELENIUM			
	Aster spp.	asters	all parts
	Astragalus bisulcatus (Hook.) Gray	milkvetch	all parts
	Astragalus pectinatus Dougl.	narrow-leaved vetch, poison vetch	all parts
	Astragalus racemosus Pursh	creamy poison vetch	all parts
	Atriplex spp.	saltbush	leaves
	Castilleja spp.	paint brush	all parts
	Comandra umbellata (L.) Nutt.	bastard toadflax	all parts
	Grindelia spp.	gumweeds	leaves
	Gutierrezia spp.	broomweeds	all parts
	Haplopappus spp.	goldenweeds	all parts
	Machaeranthera spp.	woody asters	all parts
	Penstemon spp.	beard's tongue	all parts
	Stanleya spp.	prince's plume	all parts
MISCELLANEOUS			
esculin	*Aesculus* spp.	buckeye	all parts
unknown	*Anagallis arvensis* L.	scarlet pimpernel	all parts
unknown	*Arnica montana* L.	arnica root	flowers, roots
unknown	*Artemisia filifolia* Torr.	sand sage	all parts
unknown	*Asimina triloba* (L.) Dun.	pawpaw	fruits
unknown	*Astragalus mollissimus* Torr.	woolly loco	all parts
fungus	*Avena sativa* L.	oats	smutty grain
unknown	*Cassia* spp.	senna, partridge pea	all parts
unknown	*Celastrus scadens* L.	bittersweet	leaves, bark, seeds
acrid oils	*Croton* spp.	hogwort	all parts
unknown	*Cynodon dactylon* (L.) Pers.	Bermuda grass	leaves, flowers
glycosides	*Daphne mezereum* L.	spurge laurel	all parts
unknown	*Descurainia* spp.	tansy mustard	all parts
similar to *Pteridium*	*Equisetum arvense* L.	horsetail	all parts
unknown	*Euonymus* spp.	wahoo	all parts
trematol	*Eupatorium rugosum* Houtt.	white snakeroot	all parts
unknown	*Euphorbia* spp.	spurges, poinsettia	all parts
volatile oil	*Glecoma hederacea* L.	ground ivy	all parts
gossypol	*Gossypium* spp.	cotton	seeds, meal
unknown	*Gymnocladus dioica* (L.) K. Koch	Kentucky coffee tree	leaves, seeds
trematol	*Haplopappus heterophyllus* (Gray) Blake	jimmyweed	all parts
dugaldin	*Helenium* spp.	sneezeweed, bitterweed	all parts
glycosides	*Helleborus niger* L.	Christmas rose	all parts
toxin from a fungus	*Hordeum vulgare* L.	barley	all parts
unknown	*Hyacinthus orientalis* L.	hyacinth	all parts
unknown	*Hymenoxys* spp.	bitterweed, rubberweed	all parts
unknown	*Kallstroemia parviflora* Nort.	carpet weed	all parts
lantadine A	*Lantana camara* L.	lantana	all parts
varies with species	*Lathyrus* spp.	vetch	seed
reduction of prothombin	*Lespedeza stipulacea* Maxim.	lespedeza	molded hay
unknown	*Ligustrum vulgare* L.	privet	leaves
unknown	*Lycium halimifolium* Mill.	matrimony vine	all parts
stomach irritant	*Mirabilis jalapa* L.	four o'clock	roots, seeds
unknown	*Melanthium virginicum* L.	bunch flower	all parts
unknown	*Oxytropis* spp.	locoweed	all parts
unknown	*Parthenocissus* spp.	five-leaved ivy, woodbine	fruit

Toxin	Scientific Name	Common Name	Toxic Part
amines	*Phoradendron* spp.	mistletoe	all parts
glyco-alkaloid	*Physalis* spp.	ground cherry, Jerusalem cherry, Chinese lantern	all parts
abortion	*Pinus ponderosa* Dougl.	yellow pine	needles, buds
stomach irritant	*Poinsiana gilliesii* Hook.	bird of paradise	pods
malnutrition	*Prosopis glandulosa* Torr.	mesquite	leaves, beans
unknown	*Psilostrophe* spp.	paper flower	all parts
thiaminase	*Pteridium aquilinum* (L.) Kuhn	bracken fern	all parts
several, including tannin	*Quercus* spp.	oaks	all parts
glycoside	*Rhamnus* spp.	buckthorn	all parts
robitin	*Robinia pseudoacacia* L.	black locust	all parts
unknown	*Rudbeckia laciniata* L.	goldenglow, coneflower	all parts
tanacetin	*Tanacetum vulgare* L.	tansy	all parts
unknown	*Wisteria* spp.	wisteria	pods, seeds
hydroquinone	*Xanthium* spp.	cocklebur	seeds, seedlings

SELECTED REFERENCES

Bailey, L. H. 1924. *Manual of Cultivated Plants.* Macmillan, New York.

Beath, O. A., et al. 1953. *Poisonous Plants and Livestock Poisoning.* Bulletin No. 324. Wyoming Agricultural Station, Laramie.

Choguill, H. S. 1958. "Some Poisonous Plants of Kansas." *Kansas Academy of Sciences Transactions* 61:1.

Correll, D. S., and M. C. Johnston. 1970. *Manual of the Vascular Plants of Texas.* Texas Research Foundation, Renner, Tex.

Featherly, H. I. 1945. *Some Plants Poisonous to Livestock in Oklahoma.* Circular C118. Oklahoma Agriculture Experiment Station, Stillwater.

Gates, Frank C. 1930. *Principal Poisonous Plants in Kansas.* Technical Bulletin No. 25. Agricultural Experiment Station, Kansas State University, Manhattan.

Great Plains Flora Association. 1977. *Atlas of the Flora of the Great Plains.* T. M. Barkley, editor. Iowa State University Press, Ames.

Hardin, James W. 1966. *Stock Poisoning Plants of North Carolina.* Bulletin No. 414 (rev.). Agriculture Experiment Station, North Carolina State University, Raleigh.

Hardin, James W., and Jay M. Arena. 1974. *Human Poisoning from Native and Cultivated Plants.* 2nd ed. Duke University Press, Durham, N.C.

Hulbert, L. C. and F. W. Oehme. 1968. *Plants Poisonous to Livestock.* 3rd ed. Kansas State University Printing Service, Manhattan.

Kingsbury, John M. 1964. *Poisonous Plants of the United States and Canada.* Prentice-Hall, Englewood Cliffs, N.J.

Muenscher, Walter Conrad. 1975. *Poisonous Plants of the United States.* Rev. ed. Collier Books, Macmillan, New York.

Siegmund, O. H., ed. 1973. *The Merck Veterinary Manual.* 4th ed. Merck and Co., Rahway, N.J.

Stevens, O. A. 1933. *Poisonous Plants and Plant Products.* Bulletin 265. North Dakota Agricultural Experiment Station, North Dakota State University, Fargo.

Tehon, L. R., C. C. Morrill, and Robert Graham. 1946. *Illinois Plants Poisonous to Livestock.* Circular 599. College of Agriculture Extension Service, University of Illinois, Urbana.

Trease, George, and William Charles Evans. 1972. *Pharmacognosy.* 10th ed. Baillière Tindall, London.

INDEX

159

161